SpringerBriefs in Business

SpringerBriefs present concise summaries of cutting-edge research and practical applications across a wide spectrum of fields. Featuring compact volumes of 50 to 125 pages, the series covers a range of content from professional to academic. Typical topics might include:

- A timely report of state-of-the art analytical techniques
- A bridge between new research results, as published in journal articles, and a contextual literature review
- A snapshot of a hot or emerging topic
- An in-depth case study or clinical example
- A presentation of core concepts that students must understand in order to make independent contributions

SpringerBriefs in Business showcase emerging theory, empirical research, and practical application in management, finance, entrepreneurship, marketing, operations research, and related fields, from a global author community.

Briefs are characterized by fast, global electronic dissemination, standard publishing contracts, standardized manuscript preparation and formatting guidelines, and expedited production schedules.

More information about this series at http://www.springer.com/series/8860

Kaj U. Koskinen • Rainer Breite

Uninterrupted Knowledge Creation

Process Philosophy and Autopoietic Perspectives

Springer

Kaj U. Koskinen (deceased)
Tampere University
Tampere, Finland

Rainer Breite
Tampere University
Tampere, Finland

ISSN 2191-5482 ISSN 2191-5490 (electronic)
SpringerBriefs in Business
ISBN 978-3-030-57302-7 ISBN 978-3-030-57303-4 (eBook)
https://doi.org/10.1007/978-3-030-57303-4

This Springer imprint is published by the registered company Springer Nature Switzerland AG.
The registered company address is: Gewerbestrasse 11, 6330 Cham, Switzerland

I dedicate this book to my god-daughter, Marie.

In Memory of Kaj Koskinen

Kaj retired from his post in university in 2010. Despite retiring he continued his research which was his greatest interest. When his other work at the university was left behind he was able to focus on writing. During his early years of retirement, Kaj wrote the book *Knowledge Production in Organizations* (Springer 2013), which was the third of his scientific books.

While teaching, it was characteristic of Kaj to expect his students to delve thoroughly into the subject. Sometimes, at home, he might voice aloud his annoyance at students not giving as much as he was expecting. But above all, Kaj was demanding of himself. In his writing and research, he was uncompromising. The will to understand things deeply followed him throughout his life. Coming from a modest background and respecting his father's wishes, he first studied in a vocational school and worked as a business machine technician for 10 years. With time, he was able to study further to become an engineer and later to graduate as a Master of Science. Finally in 2001, he graduated as a Doctor of Technology. According to Kaj, the last third of his working career, years 1996–2010 working in the university setting, was the happiest years of his life. Kaj had found a great passion that kept him interested for the rest of his life, and the creative process then continued after his retirement, literally until his death.

The writing of this book took two and half years. During his last years, Kaj studied philosophy at the University of Turku to get a deeper understanding of his research subject from a philosophical perspective. In this book, he wanted to extend his previous work while creating a new perspective or a way of thinking.

After being diagnosed with cancer, Kaj became worried that he might not be able to finish this book. Still, he was determined to try. Kaj continued writing unflaggingly despite increasing fatigue. The last summer of his life, he told his spouse that the only thing that managed to bring him joy anymore was writing and the thinking involved in that. He realized that there might not be a lot of time left for completing his work and accordingly got up every day to make even little progress in the writing process. Towards the end, Kaj needed help with technicalities, and he was grateful that his two adult children could help with them.

As a spouse, I came to understand that Kaj wanted to write his last book not only for the scientific community but also for his closest circle, his family, relatives, and friends. This last book Kaj dedicated to his god-daughter, Marie. I know that Kaj thought that science and independent, creative thinking were everlasting. This book is a tangible expression of that ideal and of the unconditional search for knowledge that he wanted to leave as a legacy for all he loved.

Kaj's studies identify the productive and interesting theoretical junction of the research paths, where process philosophy, the concept of knowledge, and the autopoietic point of the view meet. Together they form a fascinating research platform for those researchers who want to increase understanding of the role of the philosophical aspect of knowledge creating and sharing processes. Kaj's research findings shed light especially on autopoietic systems and their importance, when we are trying to increase understanding about the processes and systems of knowledge sharing and creating.

As Kaj has written: "Past, present and future are all right here in the present moment. The present moment is the only place where they really exist, for it is only individual awareness that places something as past, present or future. In contrast to clock time, individuals' actual experience of human time is naturally more organic than linear." Although we can examine time in different ways, we can agree that the meaning of Kaj's thoughts and research results endure in time, and they will utilize the study of knowledge now and in the future.

Päivi Kianne

Acknowledgements

I would like to extend my deepest thanks to my wife, Päivi, to my sons, Miikka and Aleksi, and to my sister, Paula, who have always greatly supported my work.

Kaj U. Koskinen

Contents

About the Authors

Kaj U. Koskinen has worked for many years as a project manager in several international engineering companies, including Outokumpu and Honeywell. Since 1997, he has been an adjunct professor in the Department of Industrial Management and Engineering at Tampere University of Technology. After retiring in 2010, he carried out studies at the Department of Philosophy at the University of Turku. Dr. Koskinen has published several international articles and books: *Autopoietic Knowledge Systems in Project-Based Companies, Knowledge Management in Project-Based Companies: An Organic Perspective* with professor emeritus Pekka Pihlanto, and *Knowledge Production in Organizations: A Processual Autopoietic View*.

Rainer Breite is working as a lecturer at Tampere University and an adjunct professor in the Turku school of Economics at the University of Turku. He was born in 1961 and worked as an engineer in several engineering companies. His main industry experience derives from water turbines. Since 1998, Dr. Breite has been a researcher, lecturer, and professor at Tampere University of Technology. His research interest is focused on firms' relationships and knowledge sharing in supply chains and networks. Dr. Breite has published several articles in journals and conferences about his research areas.

List of Figures

Chapter 1
Introduction

Humankind today is in the midst of great change, a situation that Thurow (2003) calls the third industrial revolution. It is a shift towards a knowledge-based economy, in which knowledge is the most important resource, superseding the traditional resources of land, capital and labour (Drucker 1993).

Of the theories that discuss knowledge, many focus on its contents rather than its development. Understanding knowledge contents is useful, of course, if we wish to trace the rudimentary changes in knowledge over time, an exercise that might suggest broad guidelines for knowledge development. However, a focus on contents faces severe limitations, because human knowledge is constantly changing instant by instant, sometimes through conversation and sometimes by observation. We can but briefly represent what we know before it alters.

Furthermore, many people think of knowledge as the sum of everything that we have learned. Things are generally owned by somebody, so they are property. This means that thinking of knowledge as an object leads people to focus on databases and other storage devices. They are more likely to identify legal owners of knowledge components. From this view comes the term knowledge transfer, suggesting that knowledge can be passed along like a baton in a relay race. The focus is then on identifying, organizing, collecting and measuring knowledge.

Another way of thinking of knowledge is as a process. The process perspective brings a very different focus to the domain of knowledge. People who take a process perspective focus more on the dynamic aspects of knowledge, such as sharing, creating, adapting, learning, applying and communicating. They tend to see knowledge as a dynamic soup of constantly shifting, melding and merging knowledge ingredients. They are less concerned with controlling the flow of knowledge and more interested in encouraging participation and easing communication.

K. U. Koskinen, R. Breite, *Uninterrupted Knowledge Creation*, SpringerBriefs in Business, https://doi.org/10.1007/978-3-030-57303-4_1

In this book, we concentrate on knowledge that has processual properties in its continual movement through creation, adaptation, enhancement and application.[1] Knowledge and the communication of knowledge are critical for self-sustaining systems comprised of people and the tools and machines that extend people's physical and cognitive capacities. Therefore, this book describes how process philosophy, system theory, social systems and autopoietic systems together enable uninterrupted knowledge creation.[2]

Finally, the objective of this book is to give the reader an alternative observational scheme to understand better continuous uninterrupted knowledge creation in systems. The suggested conceptual system is based on Luhmann's view of autopoiesis theory, including the process perspective, which is particularly based on the ideas of Whitehead, Rescher and their followers.

1.1 Keywords of the Book

- System (organization, like a firm)
- System theory
- Process philosophy
- Knowledge
- Autopoiesis
- Social autopoietic system
- Knowledge creation

1.2 Structure of the Book

Chapter 2 describes systems theory and systems and then briefly describes the basics of cybernetics, the systemic view and systems thinking. Then the discussion deals with the complexity of systems and organization as a system. Chapter 3 is about process philosophy. The chapter includes the notions process, time, evolution, potentiality and actuality, and change and becoming.

Chapter 4 deals with epistemology. It consists of epistemological assumptions like cognitivist epistemology, connectionist epistemology and autopoietic epistemology. Chapter 5 is about knowledge. First it describes the differences between data, information and knowledge. Then follow descriptions of processual

[1]Selected portions of this book have been previously published in Koskinen, Kaj (2013), Knowledge Production in Organizations, Springer, 978-3-319-00103-6. They are used here to update the research content, and with permission from the publisher.

[2]Selected portions of this book have been previously published in Koskinen, Kaj (2010), Autopoietic Knowledge Systems in Project-Based Companies, Palgrave, 978-1-349-32639-6. They are used here to update the research content, and with permission from the publisher.

knowledge, epistemological knowledge, individual knowledge, systemic knowledge, tacit knowledge and explicit knowledge and subsequently the notions sensemaking, meaning, experience and memory.

Chapter 6 concerns autopoiesis. The chapter consists of living systems, autopoietic systems, structural determinism and coupling an autopoietic system, and autonomy in autopoietic systems. Chapter 7 deals with social autopoietic systems: events, language, metaphors, decisions and decision makers, learning, communication, consciousness and the relation between social and psychic systems. Chapter 8 is about knowledge creation. The chapter describes the knowledge creation process (socialization, externalization, combination and internalization), two major knowledge flows (sensing as a source of major knowledge flows and memory as a source of major knowledge flows) and uninterrupted knowledge creation.

Chapter 2
Systems

In the period since the Second World War, systems theory has experienced a series of scientific revolutions or fundamental reorientations of its research perspective. This means that its research findings have thoroughly changed the concept of a system itself. Contemporary systems theory is founded on the distinction between systems and environment.

Systems take place in what may appropriately be described as a complex world, a world where there are discernible elements that are twisted together, entwined in ways that add up to an untidy mass (Hernes 2008). The mass has contours that may have names, but it is a matter of definition as to where and when one contour stops and another begins. The mass twists and unfolds continuously, which is why practitioners experience it as an unfolding process, a flow of possibilities and a conjunction of events and open-ended interactions occurring in time (Tsoukas and Chia 2005). Nevertheless, in this mass, we can identify and give names to separate strands in the form of processes, but we know to a lesser extent how the different strands interact and influence one another. Over time and space, strands mesh with other strands and together produce something of which we may sense the contours but not the full contents. Such a complex mass may be what we see as a system, knowing that it is undergoing continuous modification and reproduction.

The aim of this chapter is to explore a number of possible interpretations of the title system. Each of these selected meanings represents a theme, or a strand of study, which will be developed in one of the following chapters.

To understand how the system can deal with worrying issues arising in the strands, a view on truth and meaning is developed to promote coherent argumentation. The view is fundamentally satisfying and of general utility. This directs our interest to discourse (ideas, concepts, knowledge, etc.) and to the creators of discourse—the two being inextricably linked. By addressing the themes in terms of truth and meaning, this chapter synthesizes systems and social theory.

K. U. Koskinen, R. Breite, *Uninterrupted Knowledge Creation*, SpringerBriefs in Business, https://doi.org/10.1007/978-3-030-57303-4_2

2.1 Systems Theory and Systems

Systems theory is an interdisciplinary field of both science and the study of the systems in nature, society and science (e.g. Von Bertalanffy 1969). It is a framework within which one can analyse and/or describe any group of objects that work in concert to produce a result. This could be, for example, an organization or informational artefact. Systems theory, then, serves as a bridge for interdisciplinary dialogue between autonomous areas of study (e.g. Capra 1996; O'Connor and McDermott 1997; Mingers 2010; Luhmann 2013). The scientific research field that is engaged in the study of systems is based on the properties of systems theory and systems science. It investigates the abstract properties of the matter and organization, examining concepts and principles independent of the specific domain, substance, type or temporal scale of existence.

A system is a set of interacting or interdependent entities, real or abstract, forming an integrated whole (e.g. Anderson and Johnson 1997; Haines 1998; Jackson 2000, 2009). The concept of an integrated whole can also be stated in terms of a system embodying a set of relationships that are differentiated from the relationships of the set with other elements and from relationships between an element of the set and elements that are not part of the relational regime. Thus, the term 'system' has the following meanings (e.g. Capra 1996):

- A collection of organized things, analogous to a solar system
- A way of organizing or planning
- A whole composed of relationships among the members.

Further, following the definition by Flood (1990), a system is an abstract organizing structure that has many different paradigmatic interpretations, some of which attach systems to processes of the world while others attach systems to processes of consciousness. The main idea is of a whole characterized by richly interactive parts, and this is then expanded and/or interpreted according to various paradigms. Hence, when an individual uses the term 'system', it has two very distinct elements:

- The actual system
- The part of the system of which people are aware.

In the opinion of Jackson (2009), it is possible to identify systems of very different types:

- Physical, such as river systems
- Biological, such as living organisms
- Designed, such as automobiles
- Abstracts, such as philosophical systems
- Social, such as families
- Human activity, such as systems to ensure the quality of products.

Thus, there are many types of systems. According to Gaines (1979, p. 1), ‘A system is what is distinguished as a system’. This means that the observer has a choice regarding how to define the system that he or she intends to analyse.

Most systems share the same common characteristics. These common characteristics include, according to Von Bertalanffy (1969), the following:

- Systems have a structure that is defined by their parts and processes
- Systems are generalizations of reality
- Systems tend to function in the same way. That involves the inputs and outputs of material and then processes that cause it to change in some way
- The various parts of the system have functional as well as structural relationships. The characteristics of systems have been studied in general systems theory.

A system from this frame of reference is composed of regularly interacting or interrelating groups of activities (e.g. Kim 1999). For example, in organizations, which are complex social systems, reducing the parts from the whole reduces the overall effectiveness of organizations (Schein 1980). This is different from conventional models that centre on individuals, structures, departments and units, separate from the whole, instead of recognizing the interdependence of the groups of individuals, structures and processes that enable an organization to function. Laszlo (1972, pp. 14–15) explains that the new systemic view of organized complexity progresses ‘. . . one step beyond the Newtonian view of organized simplicity’ in reducing the parts from the whole or in understanding the whole without relation to the parts. The relationship between organizations and their environments has come to be recognized as the foremost source of complexity and interdependence. In most cases, the whole has properties that cannot be known from an analysis of the constituent elements in isolation.

There are some startling implications for the simple definition of a system. Systems function as a whole, and as a result they have properties above and beyond the properties of the parts that comprise them. These are known as emergent properties—they emerge from the system when it is working (e.g. Batterman 2001). For example, the movement of a car is an emergent property. A car needs a carburettor and a fuel tank to move, but, when an individual puts the carburettor or the fuel tank on the road, he or she sees how far it can move on its own. Properties can emerge like the beauty of a rainbow when the rain, atmosphere and angle of sunlight fit together perfectly. Because people live with emergent properties, they take them for granted, yet they are often unpredictable and surprising.

2.2 Cybernetics, the Systemic View and Systems Thinking

Cybernetics is a trans-disciplinary approach to exploring regulatory systems—their structures, constraints and possibilities (e.g. Jackson 2003). The term is often used in a rather loose way to imply ‘control of any system using technology’. In other words, it is the scientific study of how humans, animals and machines control and communicate with each other.

Furthermore, cybernetics is the study of feedback, and it is derived from concepts such as communication and control in living organisms, machines and organizations. Its focus is on how an object (digital, mechanical or biological) processes data, reacts to data and changes or can be changed to accomplish the first two tasks better (e.g. Wiener 1948).

The terms systems theory and cybernetics are widely used as synonyms. Some authors use the term cybernetic systems to denote a proper subject of the class of general systems, namely those systems that include feedback loops. Cybernetics, catastrophe theory (e.g. Gilmore 1981), chaos theory (e.g. Levy 1994) and complexity theory (e.g. Mitleton-Kelly 2003) have a common goal, which is to explain complex systems that consist of a large number of mutually interacting and interrelated parts in terms of those interactions.

Cybernetics is applicable when a system being analysed incorporates a closed signalling loop—originally referred to as a 'circular causal' relationship—that is, one in which action by the system generates some change in its environment and that change is reflected in the system in some manner (feedback) that triggers a system change. Cybernetics is relevant to, for example, mechanical, physical, biological, cognitive and social systems. The essential goal of the broad field of cybernetics is to understand and define the functions and processes of systems that have goals and that participate in circular, causal chains that move from action to sensing to comparison with the desired goal and again to action. Its focus is how anything (digital, mechanical or biological) processes information, reacts to information and changes or can be changed to accomplish the first two tasks better. Thus, cybernetics includes the study of feedback, black boxes and derived concepts such as communication and control in living organisms, machines and organizations, including self-organization (Jackson 2009).

All in all, the concepts studied by cyberneticists include, but are not limited to, learning, cognition, adaptation, social control, emergence, convergence, communication, efficiency, efficacy and connectivity. In cybernetics, these concepts (otherwise already objects of study in other disciplines, such as biology and engineering) are abstracted from the context of the specific organism or device.

The systemic view is the view that all systems are composed of interrelated subsystems (Koskinen 2013). A whole is not just the sum of the parts, but the system itself can be explained only as a totality. The systemic view is, therefore, the opposite of reductionism, which views the total as the sum of its individual parts. In traditional organization theory, as well as in many of the sciences, the subsystems are studied separately, with a view to putting the parts together into a whole at some later point. The systemic view emphasizes that this is not possible and that the starting point has to be the total system (Koskinen 2013). When people look at the patterns that connect the parts rather than simply the parts themselves, a remarkable fact emerges. Systems made from very different parts that have completely different functions follow the same general rules of organization. Their behaviour depends on how the parts are connected rather than what the parts are. Therefore, one can make predictions about their behaviour without knowing the parts in detail. That is, one can understand and influence very different systems using the same principles.

Furthermore, the systemic view of organizations is transdisciplinary and integrative. This view transcends the perspectives of individual disciplines, integrating them on the basis of a common code, that is, on the basis of the formal apparatus provided by systems theory (e.g. Haines 1998; Bell and Morse 1999; Jackson 2009). The systemic view gives primacy to the interrelationships rather than to the elements of a system. It is from these dynamic interrelationships that new properties of the system emerge. As Capra (1996) argues, the more we study the major problems of our time, the more we come to realize that they cannot be understood in isolation. They are systemic problems, which means that they are interconnected and interdependent.

Systems thinking comes from a rigorous scientific discipline called general system theory (e.g. Von Bertalanffy 1969; Weinberg 2001), which developed from the study of biology in the 1920s. The theory centred on the natural world, the living systems therein and the common laws governing those systems (Haines 1998). Its major premise was that such laws, once known, could serve as a conceptual framework for understanding the relationships within any system and for handling any problems or changes encompassed by that system. Consequently, the theory emphasized the value of viewing a system as a whole, of gaining a perspective on the entire entity before examining its parts. It is this emphasis that informs and shapes the practice of systems thinking.

Furthermore, systems thinking means a shift in attention from the part to the whole (Checkland 1997; Weinberg 2001; Jackson 2003), considering the observed reality as the integration and interaction of phenomena whereby the individual properties of the single parts become indistinct. In contrast, the relationships between the parts themselves and the events that they produce through their interaction become much more important, with the result that the system elements are rationally connected (Luhmann 1990) and directed towards a shared purpose (Golinelli 2009).

Then, according to Anderson and Johnson (1997), systems thinking can be characterized as follows:

- Thinking of the 'big picture'
- Balancing short-term and long-term perspectives
- Recognizing the dynamic, complex and interdependent nature of systems
- Taking into account both measurable and non-measurable factors
- Remembering that we are all part of the systems in which we function and that we each influence those systems even as we are being influenced by them.

Moreover, systems thinking is a basis for clear thought and communication, a way of seeing more and further (e.g. O'Connor and McDermott 1997; Mingers 2010). This means that obvious explanations and majority views are not always right. With a wider and different perspective, an individual can see exactly what is happening and then take the actions that he or she knows are the best in the long run. Systems thinking thus considers the whole, the parts and the connections between the parts, studying the whole to understand the parts.

In practice, systems thinking is any process of estimating or inferring how actions or changes influence the state of the neighbouring systems. It is an approach to problem solving that views problems as parts of an overall system rather than reacting to present outcomes or events and potentially contributing to further development of the undesired issue or problem. In other words, systems thinking is a framework that is based on the belief that the components of a system can best be understood in the context of their relationships with each other and with other systems rather than in isolation. The only way to understand fully why a problem or element occurs and persists is to understand the part in relation to the whole. This means that systems thinking is a way to view and frame mentally what we see in the world, that is, a worldview and way of thinking whereby we see the entity or unit first as a whole, with its fit and relationship with its environment as primary concerns.

Furthermore, according to Capra (1996), systems cannot be understood by analysis. The properties of the parts are not intrinsic properties but can be understood only within the context of the larger whole. Thus, the relationship between the parts and the whole has been reversed. In the systems approach, the properties of the parts can be understood only from the organization of the whole. Accordingly, systems thinking concentrates not on the basic building blocks but on the basic principles of organization. Systems thinking is then contextual, which is the opposite to analytical thinking. Analysis means taking something apart to understand it; systems thinking means putting it into the context of a larger whole.

The reason that habitual thinking is insufficient to deal with systems is because it tends to see simple sequences of cause and effect that are limited in time and space rather than a combination of factors that mutually influence each other. In a system, cause and effect may be far apart in time and space. The effect may not be apparent until days, weeks or even years later. Still, people have to act now (O'Connor and McDermott 1997). In other words, high-leverage changes are usually not obvious to most participants in the system. They are not close in time and space to obvious problem symptoms. Systems thinking, however, shows that small, well-focused actions can sometimes produce significant, enduring improvements if they are in the right place, through leverage (cf. Marquardt 1996, p. 43).

In sum, the systemic view and systems thinking are based on a simple but profound truth: 'Living systems are the natural order of life' (Haines 1998, p. viii). Thus, the systemic view and systems thinking are alternatives to the pervasive reductionism in Western culture—the pursuit of simple answers to complex issues. Then, the systemic view and systems thinking attempt to illustrate that events are separated by distance and time and that small catalytic events can cause large changes in systems, acknowledging that an improvement in one area of a system can adversely affect another area of the system, which promotes organizational communication at all levels to avoid the silo effect. Systems thinking has been developed to provide techniques for studying systems in holistic ways to supplement the traditional reductionist methods. In this more recent tradition, the systemic view in organizational studies is considered by some as a humanistic extension of the natural sciences.

2.3 Complexity of Systems

The complexity of organizations is often discussed in the organization literature without a clear definition of the concept (Luhmann 1990). Mostly, the term 'complexity' is contrasted with the term 'simplicity'. The underlying distinction is that between a whole and its parts. A unity, in this sense, is complex if it is composed of several parts and simple if it cannot be decomposed further (Luhmann 1992, p. 364).

Traditionally, the term 'complexity' could be defined through a distinction, one side of which contained the complex and other the simple. The distinction marks mutually exclusive terms: something has to be either simple or complex (Luhmann 1993, p. 64). This definition, however, proves to be inconsistent, since all simple phenomena can be shown to be themselves composed of parts and thus complex. Especially from physics, we learn that there are no simple phenomena; everything that is taken as an element (atoms, particles, quarks) is just a temporary limit of further decomposition (Luhmann 1993, p. 61). Thus, because all simple phenomena are themselves complex, the two terms cannot be mutually exclusive. The distinction complex/simple contains complexity on both sides and thus does not help to define complexity.

Instead of basing the concept of complexity on the distinction between a whole and its parts (i.e. simple parts making up a complex whole), Luhmann suggests basing it on the distinction between element and relation. Something is considered to be complex if there are more possible relations between its elements that can be realized. For example, a system is complex if it cannot relate every element to every other element (Luhmann 1995a, p. 24).

That is, complexity refers to the phenomenon whereby there are more elements than can be related to each other. In this sense, the system has higher potential than it can actualize. Hence, instead of defining complexity as distinct from simplicity, Luhmann suggests treating complexity itself as a distinction: the distinction between complete and selective relationing between elements.

2.4 Organization as a System

Recently, management studies have come to view organizations from a new perspective: a systems perspective. This systems perspective may seem quite basic, yet decades of management training and practices in the workplace have not adopted this perspective. Only recently, with tremendous changes facing organizations and the way in which they operate, have educators and managers come to face this new way of looking at things. This interpretation has brought about a significant change (or paradigm shift) in the way in which management studies and approaches organizations.

The effect of this systems theory in management is that writers, educators, consultants and so on are helping managers to view organizations from a broader perspective. Systems theory has brought a new perspective for managers to interpret patterns and events in their organizations. In the past, managers typically took one part and focused on that. Then they moved all their attention to another part. The problem was that an organization could, for example, have wonderful departments that operated well by themselves but did not integrate well together. Consequently, the organization suffered as a whole.

Now, more managers are recognizing the various parts of the organization and, in particular, the interrelation of the parts, for example the coordination of central offices with other departments, engineering with manufacturing, supervisors with workers and so on. Managers now focus more attention on matters of ongoing organization and feedback. Managers now diagnose problems not by examining what appear to be separate pieces of the organization but by recognizing larger patterns of interactions. Managers maintain their perspective by focusing on the outcomes that they want from their organization. Now managers focus on structures that provoke behaviours that determine events—rather than reacting to events, as always occurred in the past.

Thus, an organizational system is the structure of an organization's set-up. That structure defines how each element of a system is set up and to which element and how communication flows throughout the organization. With a well-defined organizational structure in place, all the elements (e.g. individuals) know what is expected of them and to whom they report. An organizational structure that is right for one system will not be right for another. Hence, in this book, different organizations (e.g. business firms, sport clubs, etc.) are seen as systems, and therefore the term 'system' is used to name them. However, it is very important to realize that, when the discourse deals with the autopoietic elements, then 'organization' is the right term.

2.5 Summary

In this chapter, we have covered a considerable amount of ground, looking at the different approaches and terms of systems theory. The systems approaches available today have resulted from attempts to correct the original problems found when trying to use the systemic view in practice. They have also arisen from theoretical developments in the transdiscipline of systems thinking as new problem contexts have been envisioned and their implications for practice have been explored. As seen above, it is reasonable to conclude that there are numerous strands of systems and concepts embracing a whole variety of individual systems approaches. Furthermore, today people have learned to adopt a systems perspective. They work to improve their systems not by examining what appear to be separate pieces of the system but by recognizing the larger patterns in interactions.

Chapter 3
Process Philosophy

The progenitor of this metaphysical tradition was Heracleitus. For him, reality is not a constellation of things at all but one of processes. The fundamental 'stuff' of the world is not material substance but volatile flux, namely 'fire', and all things are versions thereof. Process is fundamental: the river is not an object but a continuing of flow; the sun is not a thing but an enduring fire. Everything is a matter of process, of activity, of change. It is not stable things but fundamental forces and the varied and fluctuating activities that they manifest that constitute the world (e.g. Rescher 2000).

Process philosophy has two closely interrelated sectors, one epistemic and the other ontological. In other words, process philosophy centres on metaphysics and ontology, but it has full systematic scope: its concern is with the dynamic sense of being as becoming or occurrence, the conditions of spatiotemporal existence, the kinds of dynamic entities, the relationship between mind and world and the realization of values in action. Some approaches to process philosophy are conceived on a grand scale and offer full-scope metaphysics in the form of a systemic theory or comprehensive philosophical view.

The epistemic aspect is based on the idea that a process and its ramifications afford the most appropriate and effective conceptual instruments for understanding the world in which we live. Thus, the epistemic aspect is an approach to epistemology that emphasizes the truth-conduciveness of a belief-forming process, method or other epistemologically relevant factor (e.g. Rescher 2004). The reliability theme appears in theories of knowledge, of justification and of evidence. The epistemic aspect is sometimes used broadly to refer to any theory that emphasizes truth-gaining or truth-indicating properties. More commonly, it is used narrowly to refer to a process for justification. This entry discusses truth in both broad and narrow senses but concentrates on the theory of justification.

The ontology aspect refers to a universal model of the structure of the world as an ordered wholeness (Rescher 2004). Such an ontology is in contrast to the so-called applied ontology. The ontology-type process philosophy does not claim to be

K. U. Koskinen, R. Breite, *Uninterrupted Knowledge Creation*, SpringerBriefs in Business, https://doi.org/10.1007/978-3-030-57303-4_3

accessible to any empirical proof in itself but to be a structural design pattern, through which empirical phenomena can be explained and put together consistently.

According to Whitehead (1938, p. 71), there are three ways of categorizing human experience: by substance, happenings and 'absolute'. The first implies a view whereby there are enduring substances out there to which phenomena can be related. However, Whitehead rejects that idea, arguing that substances have no place in existence. What we see as substances are processes, and the only real units of analysis are the happenings of experience. The third category (i.e. absolute) belongs to the question of God and will not be pursued in this book. As for the first two categories, Whitehead insists that the choice stands between two ultimate sides: 'One side makes process ultimate; the other side makes fact ultimate' (Whitehead 1978, p. 7).

Thus, according to Whitehead (1978), the world cannot be anything but process. Hence, there can be no static entities on which we can base our thinking, because everything is in a state of becoming. The real world is in a continuous state of flux, which is infinitely complex and cannot be understood as such. However, parts of it may be sensed. It is from this indecipherable world of flux that understanding, represented by models of the world, may be extracted and made objective and logical (Hernes 2008).

In the opinion of Rescher (1996), the characteristic feature of process philosophy is its stress on the primacy of activity and on the range of associated factors, such as time, change, innovation and so forth. It maintains that these conceptions are not just necessary but even basic to our understanding of the world. Process philosophy does not deny things but sees them as subordinate in status and ultimately inherent in processes. Then, process philosophy stresses the reality and prominence of literally autonomous, altogether un-owned processes that do not merely represent the activities of substances. A fundamental feature of the real is that it is processual in nature, and ultimately the only satisfactory way to understand what things are is to proceed in terms of what they do.

Furthermore, process philosophy is a longstanding philosophical tradition that emphasizes becoming and changing over static being. Process philosophy is characterized by an attempt to reconcile the diverse intuitions found in human experience, such as religious, scientific and aesthetic, into a coherent holistic scheme. Process philosophy seeks a return to a neo-classical realism that avoids subjectivism. This reconciliation of the intuitions of objectivity and subjectivity, with a concern for scientific findings, produces the explicitly metaphysical speculation that the world, at its most fundamental level, is made up of momentary events of experience rather than enduring material substances. Process philosophy speculates that these momentary events, called 'actual entities' or 'actual occasions' (Sherburne 1966; Whitehead 1978), are essentially self-determining, experiential and internally related to each other.

Thus, according to Whitehead (1978), the world cannot be anything but process. Hence, there can be no static entities on which people can base their thinking, because everything is in a state of becoming. The real world is in a continuous state of flux, which is infinitely complex and cannot be understood as such.

However, parts of it may be sensed. It is from this indecipherable world of flux that understanding, represented by models of the world, may be extracted and made objective and logical (Hernes 2008).

Moreover, process philosophy identifies metaphysical reality with change and development. Since the time of Plato and Aristotle, philosophers have posited true reality as timeless, based on permanent substances, while processes are denied or subordinated to timeless substances (e.g. Langley and Tsoukas 2010). For example, if Socrates changes, becoming sick, Socrates is still the same, and his sickness only glides over his substance. Then, the change is accidental, whereas the substance is essential. That is why classical ontology denies any full reality to change, which is conceived as only accidental and not essential. This classical ontology is what made knowledge and theory of knowledge possible, as it was thought that a science of something in becoming was an impossible feat to achieve (Fagot-Largeau 2006).

In opposition to the classical model of change as accidental or illusory, process philosophy regards change as the cornerstone of reality—the cornerstone of being thought of as becoming. In physics, for example, Prigogine (1980) distinguishes between the 'physics of being' and the 'physics of becoming'. Process philosophy covers not just scientific intuitions and experiences but can be used as a conceptual bridge to facilitate discussions among religion, philosophy and science (Weber 2004; Hustwit 2007). Other approaches take a more modest approach. They pursue the specific problems that the various philosophical disciplines are engaged in while focusing on the dynamic aspects of each sub-domain. In other words, process philosophy is based on the premise that being is dynamic and that the dynamic nature of being should be the primary focus of any comprehensive philosophical account of reality and our place within it.

In the opinion of Mesle (2008), process philosophy is a vision that matters, that is, worth taking seriously, even though it will inevitably require people to criticize, improve and, in some respects, transcend it. Anyway, process philosophy is an effort to think clearly and deeply about the obvious truth that the world and the lives of individuals are dynamic, interrelated processes and to challenge the apparently obvious, but fundamentally mistaken, idea that the world is made of things that exist independently of such relationships and that seem to endure unchanged through all the processes of change.

According to Rescher (2000), it seems sensible to understand process philosophy as a doctrine that is committed to, or at any rate inclined towards, certain basic propositions:

1. Time and change are among the principal categories of metaphysical understanding.
2. Process is a principal category of ontological description.
3. Processes are more fundamental, or at any rate not less fundamental, than things for the purposes of ontological theory.
4. Several, if not all, of the major elements of the ontological repertoire (God, nature as a whole, persons and material substances) are best understood in process terms.

5. Contingency, emergence, novelty and creativity are among the fundamental categories of metaphysical understanding.

To sum up, in recent years, process philosophy has become a catchphrase for the doctrines of Alfred North Whitehead and his followers. However, this cannot really be what process philosophy ultimately concerns. If a philosophy of process exists, it must pivot not on a thinker but on a theory. What is at issue must, in the end, be a philosophical position that has a life of its own, apart from any particular exposition or expositor.

Any fair-minded and conscientious view of the matter must acknowledge that process philosophy is a complex and multifaceted project that resists any attempt to constrain it neatly and narrowly in the pre-established philosophical textbook typology. Process philosophy is so many-sided that it abuts on every area of philosophical concern.

Moreover, all the systems can be seen as different aspects of one great strand of systemic thinking, which may be called contextual thinking. There is another strand of equal importance, which emerged somewhat later in twentieth-century science. This second strand is process thinking. In the mechanistic framework of Cartesian science, there are fundamental structures, and then there are forces and mechanisms through which these interact, thus giving rise to processes. In systems science, every structure is seen as the manifestation of underlying processes. Systems thinking is always process thinking. Most fundamentally, process thinking is based on the worldview that sees processes rather than substances as the basic forms of the universe (Bergson 1946; Whitehead 1978; James 1996).

All in all, process philosophy emphasizes the elements of becoming, change and novelty in experienced reality; it opposes the traditional philosophical stress on being, permanence and uniformity.

3.1 Process

Process is a generic term for the continuous making and moving of forms (e.g. Cooper 2006). The dictionary defines a process as a series of acts or events. An act is a physical and mental gesture that constitutes the actual making of a form, while an event is an act that has yet to complete itself as an enduring presence. A process in this sense reminds us that the ready-made structures of the human world are initiated and constituted by human agency and are thus not objectified forms independent of the human observer. Observer and observed are mutually constituting acts of knowing that generate each other. A process has to be understood as the continuous coming-to-presence of the forms and objects of everyday life rather than their taken-for-granted, ready-made presences. The act of observation necessarily requires the active participation of the human agent as an observer in making the world present and presentable in repeated acts of representation. Reality in this sense is realization of the world as a source of appearances and forms rather than their objective, independent existences. This is what the origin of process implies in its etymological combination of approach and withdrawal as a process. A process can thus be understood as a divided state of being in which human agency is forever suspended between the ceaseless act of making forms present and their constant recession. Process and recess are recursive versions of each other in a world without end.

The conception of a process is complex. It combines the features of internal make-up with change and development in time. A process is an actual or possible occurrence that consists of an integrated series of connected developments unfolding in programmatic coordination: an orchestrated series of occurrences that are systematically linked to one another either causally or functionally. That is, everything is in process, becoming and perishing (Rescher 2000).

A natural process by its very nature passes on to the future a construction made from the materials of the past. In other words, all processes have a developmental, forward-looking aspect. Each such process envisions some sector of the future and canalizes it into regions of possibility that are more restrained in range than would otherwise, in theory, be available. The inherent future of process is an exfoliation of the real by successively actualizing possibilities that are subsequently left behind as the process unfolds. Ultimately, it is a question of priority of viewing the time-bound aspects of the real as constituting its most characteristic and significant features. This is to say, process has priority over product—both ontologically and epistemically.

According to Rescher (1996), in general, processes have three phases or stages: initiating precursors, the process itself and resultant successors. Processes exhibit such a pattern of sequential order, be it a temporal or a conceptual order that is at issue.

Processes develop over time: any particular natural process combines existence in the present with tentacles that reach into the past and the future. Even cognitive processes vary with the passage of time. Processes will always involve a variety of subordinate processes and events, just as creating a scientific article involves its writing, publishing and distribution.

All in all, a process is a complex of occurrences—a unity of distinct states or phases. A process is always a matter of now this, now that (Rescher 2000). This complex of occurrences has a certain temporal coherence and integrity, and processes accordingly have an ineliminably temporal dimension. A process has a structure, a formal generic patterning of occurrence, through which its temporal phases exhibit a fixed format.

3.2 Time

There is a general agreement among philosophers that time is continuous (i.e. we do not experience it as stopping and starting or darting about at random) and that it has an intrinsic direction or order (i.e. we all agree that events progress from past to present to future) (e.g. Mesle 2008). There is also a more or less general agreement that it is objective and not subjective or dependent on its being consciously experienced, which is borne out by the almost universal agreement on the time order of so many events, both psychological and physical, and the fact that so many different physical processes bear consistent time relations to each other (e.g. the rotation of the earth, the frequency of oscillation of a pendulum, etc.). However, even given that, many differing opinions and approaches to what time actually is have been put forward over the centuries.

Furthermore, time has long been an important subject of study in religion, philosophy and science, but a definition that is applicable to all fields without circularity has consistently eluded scholars. Nevertheless, diverse fields such as business, industry, sports, the sciences and the performing arts all incorporate some notion of time into their respective measuring systems.

The traditional view of time is as clock time, in which past, present and future are all separate and distinct from each other. In the opinion of Allee (1997), clock time is a hard task mistress. Individuals' relationship with clock time is adversarial, as they struggle to cram their activities into its narrow confines. However, the clock time approach negates human time, in which, in actual experience, the past, present and future are all inter-related (Savage 1996). Past, present and future are all right here in the present moment. The present moment is the only place where they really exist, for it is only individual awareness that places something as past, present or future. In contrast to clock time, individuals' actual experience of human time is naturally more organic than linear. However, people have conditioned themselves so strongly by the clock that they seldom experience their natural time rhythms. Even individuals' workspaces insulate them from the rhythm of nature, the seasons and the turn of the day.

According to Luhmann (1995b), time forces us to organize. Time is a precious restraint that impels us to organize so that we can achieve things. Luhmann (1995b, p. 42) points out that, if an infinite amount of time was at one's disposal, everything could be brought into tune with everything else. A consequence of an abundance of time would be that we would not be forced to make selections, but, because time is in limited supply, we are forced to select the things that we think we can control and select away things that are not within our sphere of control or our sphere of understanding. This is how we construct our world of relative stability in a world where everything flows (Chia 2000).

Thinking about process is about coming to grips with the phenomenon of being in the flow of time (Hernes 2014). It is not about the flow of things but about the things of flow. The things of flow include the actors who find themselves having to organize the world around them in the flow of time. They cannot simply step out of the flow, decide how to organize others and then step back into the flow, because such an ideal scenario would presume that one could stop time, or even reverse it, and start it again. The point is that even freezing time, by going on leave, by taking a holiday or, very unlikely, by ordering everyone to do nothing and turn off all machines, does not stop the flow of time. Attempting to substitute that which is with emptiness does not put an end to what is happening. Instead, the actors have to organize while on the go, while, in the flow of time, staying entangled with everything else that makes up the flow.

A modern philosophical theory called presentatism views the past and the future as human mind interpretations of movement instead of real parts of time (or dimensions), which coexist with the present. This theory rejects the existence of all direct interactions with the past or the future, holding only the present as tangible. This is one of the philosophical arguments against time travel. This contrasts with eternality (all time, present, past and future, is real) and the growing block theory (the present and the past are real, but the future is not).

Whitehead (1978, pp. 126, 196) describes time as perpetually perishing, by which he means that the experience of particular duration is forever perishing. It comes into being and then vanishes. It never really is, except in the imagination. Continuity, therefore, is an imagined state of affairs over duration. Whereas passage may be seen as the ontology of becoming, its epistemological remedy consists of imagining continuity and, with it, change. This is why rituals hold such enormous importance in systemic life (March and Olsen 1989) and why they become more important the more volatile life appears to be.

Inherent in each actual entity is its respective dimension of time. Potentially, each Whitehead an occasion of experience is causally consequential on every other occasion of experience that precedes it in time and has as its causal consequences every other occasion of experience that follows it in time; thus, it has been said that Whitehead's occasions of experience are all windows, in contrast to Leibniz's windowless monads. In time defined relative to it, each occasion of experience is causally influenced by prior occasions of experiences and causally influences future occasions of experience. An occasion of experience consists of a process of prehension with other occasions of experience, reacting to them. This is the process in process philosophy. Such a process is never deterministic. Consequently, free will is essential and inherent to the universe.

The causal outcomes obey the usual well-respected rule that the causes precede the effects in time. Some pairs of processes cannot be connected by cause-and-effect relations, and they are said to be spatially separated. Time in this view is relative to an inertial reference frame, different reference frames defining different versions of time.

To sum up, flow includes the actors who find themselves having to organize the world around them in the flow of time. In other words, actors have to organize while in the flow of time, staying entangled with everything else that makes up the flow.

A traditional view of time is as clock time. However, clock time negates human time. Giving an active role to time is particularly important given the present evolution of systemic life. Time matters in a dual sense. It matters because it is more important than ever, as the world is apparently continually being sucked into a whirl of crises and events. Time is also of significance because it gives meaning to matter.

3.3 Evolution

The scientific theory of evolution by natural selection was proposed by Charles Darwin and Alfred Russel Wallace in the mid-nineteenth century and was set out in detail in Darwin's book *On the Origin of Species* (1859). Evolution by natural selection was first demonstrated by the observation that more offspring are often produced than can possibly survive. This is followed by three observable facts about living organisms:

- Traits vary among individuals with respect to their morphology, physiology and behaviour (phenotypic variation)
- Different traits confer different rates of survival and reproduction (differential fitness)
- Traits can be passed from generation to generation (heritability of fitness). Thus, in successive generations, members of a population are more likely to be replaced by the progeny of parents with favourable characteristics that have enabled them to survive and reproduce in their respective environments.

Thus, evolutionary theory is the area that focuses on further development and refinement of the modern synthesis of evolution and genetics. Notable topics include the appropriate level of selection, the relative importance of natural selection and other mechanisms and the rate of evolution at the genotypic and phenotypic levels.

Furthermore, evolutionary theories are a class of theories, models or arguments that explain how systems evolve and why successful systems like business firms differ from each other. They explain the generation and renewal of variation by random elements and winnowing. Internal forces provide continuity to whatever survives the winnowing. Many of the evolutionary theories assume that individual learning, systemic adaptation and environmental selection of systems are taking place at the same time (Nelson and Winter 1982; Nelson 1994, 1995).

An important aspect of the classical theory of evolution is the idea that, in the course of evolutionary change and under the pressure of natural selection, organisms will gradually adapt to their environment until they reach a fit that is good enough for survival and reproduction. However, in the new systemic view, evolutionary change is seen as the result of life's inherent tendency to create novelty, which may or may not be accompanied by adaptation to changing environmental conditions.

Evolutionary theories can also be regarded as learning theories (Dodgson 1993). Foss et al. (1995) attempt to explain technological evolution and competition through a set of variables that change over time as well as the dynamic process behind the observed change. These theories are process oriented and based on routines that preserve and stabilize systemic behaviour. They focus primarily on intangible resources, whereas the resource-based theory focuses in principle on all resources.

Moreover, evolutionary theories are consistent with the Schumpeterian evolutionary view of economic process and change. They focus on the dynamic process of social construction and on the transformation of alternative forms within and across generations of competing systemic routines, forms and institutions (Nelson 1994). Evolutionary theories are explanatory process theories, not predictive ones (van de Ven 1992). Their level of analysis has conventionally been an industry, and the main emphasis has been on system populations. However, Foss et al. (1995) do not agree with this view, and, according to them, an evolutionary theory of the firm has largely been lacking. Fortunately, the new evolutionary literature is sensitive to intra-system, system, population and community evolution (Baum and Singh 1994a, b; Aldrich 1999).

Baum and Singh (1994a) write that, since the 1960s, the open-system model, in which the environment is located outside the system, has been the prominent view of system theory. However, the environment can be treated as exogenous only if the system of variables is in equilibrium. In different conditions, it is more useful to take a co-evolutionary approach and view each variable as influencing the others.

To sum up, it is natural to think of the history of systems in evolutionary terms, as each system competes with the others for scarce resources and their fates must consequently be decided by some combination of natural selection and rational adaptation (e.g. Simon 1993). Evolution is an emblematic and paradigmatic process for process philosophy. Not only is evolution a process that makes process philosophy possible, but also it provides a clear model for how processual novelty comes into operation in nature's self-engendering and self-perpetuating scheme of things. Evolution, be it of organism or of mind, of subatomic matter or of the cosmos as a whole, reflects the pervasive role of process. Change pervades nature. The passage of time leaves neither individuals nor types (species) of things statically invariant. Process at once destabilizes the world and is the cutting edge of advance to novelty. Evolution of every level, physical, biological and cosmic, carries the burden of the work here.

3.4 Potentiality and Actuality

Events, according to Whitehead (1938, pp. 99–100), are intimately connected to potentiality and actuality. This means that events drive processes by transforming actuality into potentiality.

The potentiality–actuality dimension is a general principle of processes. The atomistic view held by Whitehead offers a view whereby one form of a system emerges among several possible forms. The form that emerges is the actual form, emerging as one of many different possibilities. The actual form of a system consists of sets of interconnected abstractions that provide meaning to actors, both within and outside the system. That is, according to Hernes (2008), there are many ways in which the form that exists is derived from potentialities of previous forms. At the same time, the actual form represents the potentiality for possible future actual forms. The potentiality provides a number of factors within the reach of actors and outside it and holds potential for causing unanticipated consequences.

A system implies reiteration of the connections between entities in the making while, at the same time, harbouring possibilities for connecting with other entities. As systems become organized, some opportunities reveal themselves through connections while others are omitted. Thus, systems are continually faced with the necessity of making selections, which also means selecting away possibilities. In fact, selection makes no sense without an awareness of what is selected away. Selecting something over something else means drawing a distinction between the selected and the non-selected.

Potentiality, as opposed to actuality, represents that which cannot be accounted for; it represents that which is absent, that which is not available for assessment, but which nevertheless represents a space of opportunities. Potentiality resides in the unintended consequences of what actors do, but it also resides in forces in the environment. It is important to keep in mind that potentiality works in two ways in relation to actuality. On the one hand, actuality creates potentiality in the sense that what is acted has effects on what happens later, beyond that which was anticipated. This is inevitable in organizing processes. On the other hand, actuality embodies potentiality from earlier actualities and from actualities elsewhere.

To sum up, in process philosophy, potentiality and actuality constitute a pair of closely connected principles, which is used to analyse motion, causality, ethics and physiology. The idea of potentiality holds that everything existing in time–space as actuality holds potentiality for actual experience elsewhere in time–space. That is, seemingly disconnected events may hold potential effects on each other. Using an example given by Hernes (2008), a technology developed at one end of the world may influence greatly what people do at the other end of the world, and it may consequently influence the social identity of the same people, which may change through their adoption of the technology.

3.5 Change and Becoming

Systems are commonly conceptualized as entities adapting to the environment, and analysis consequently becomes focused on the work of adapting to an environment that changes between successive stable states. Therefore, systemic change has commonly been seen as a stepwise adaptation to changes in the environment, and consequently analysis loses out on the possibility of understanding the dynamics by which systemic actors work to make a difference. From a process perspective, the focus is inverse: it is stabilization, and not change, that needs to be explained, because the world is continually changing and a system consists of attempts at stabilization to create a predictable world amid multiple possibilities. However, the work of stabilization is fraught with uncertainty and ambiguity. It demands that actors envisage what may actually become while preparing for various future potentialities rather than adapting to some externally given quasi-stability of affairs.

However, according to Tsoukas (2005), people do not know enough about how change is actually accomplished. Even if they can explain how and why a system moved from one position to another, the explanation would not be detailed enough to show how change was actually accomplished on the ground, how plans were translated into action and, in so doing, how they were modified, adapted and changed. Feldman (2000, p. 613) empirically shows how systemic routines, far from being the repeated stable patterns of behaviour that do not change very much from one iteration to another, are actually emergent accomplishments that perpetually interact and change in action. According to Tsoukas (2005), in so far as routines are performed by human agents, they contain the seeds of change.

In the opinion of Hernes (2014), a common way to distinguish process thinking from other approaches is to associate it with the becoming of things. The idea of using the word becoming is to indicate explicitly that things are not to be considered as existing in a final state and that whatever entity we are considering (a human being, a machine, a routine or a goal) is in a continuous state of becoming through the work of connecting it with other things. This is the very basis of a relationship, as expressed by process sociologists such as Elias (1978) and Emirbayer (1997).

According to some process-oriented thinkers (e.g. Dawson 1994; Hernes 2008), change must not be thought of as a property of a system. Rather, a system must be understood as an emergent property of change. Change is ontologically prior to a system; it is the condition of possibility for a system and systemic identity. Drawing on process-oriented philosophers and ethnomethodologists, Tsoukas and Chia (2005) argue that change is the reweaving of individuals' webs of beliefs and habits of action as a result of new experiences obtained through interactions. This is an ongoing process in which individuals try to make sense of, and act coherently in, the world. Thus, change is inherent in human action.

The idea of becoming should not be seen as becoming else or the successive becoming of something different, as that could suggest a continuously changing entity, and the question would be whether it is changing and how much. A major implication is that becoming is an experience in time and that the only reality that matters is the experience of the entity in question.

Continuity offers opportunity for change. According to the principle of autopoiesis, systems uphold themselves through interaction with their own states. A system breaks down without reproduction. On the other hand, if there is only reproduction of the existing features of the system, the system cannot change and will remain essentially identical over time. The dilemma is, according to Hernes (2008), resolved by conceptualizing the relationship between process and structure. A process, consisting of successive events, offers occasions for change as well as continuity (Luhmann 1995a, p. 347). Change may happen in a number of ways (e.g. through accidents or unintended consequences), but it will take hold only insofar as it can be understood by the system. In other words, it is interpreted through the codes of communication that are appropriate to the system in question. A structure presupposes self-maintenance, which is sufficiently stable to enable meaning to be made of opportunities for change, thus enabling choices to be made against a horizon of recognizable possibilities.

To sum up, an entity is not to be taken as an individual person or social unit but as a relational whole, which may be taken as a system (e.g. Whitehead 1978). An important contribution from process philosophers is the idea of explaining how processes make the subject rather than assuming that the subject exists prior to the processes while granting primacy to time. What sometimes distinguishes process philosophers from sociologists is that the notion of the process is not confined to social processes but is meant in a broader, sometimes metaphysical, sense. This is why the idea of the relational whole is important.

3.6 Summary

To summarize, one way to deal with the stability versus change conundrum is to replace stability with continuity and accept that there is continuity in change, which implies taking a temporal view of systemic life. A temporal view implies coming to grips with how continuity and change become understood with the passing of time. From a process philosophical outlook, primacy is accorded to becoming over being, difference over self-identity and time and temporality over simple spatial location.

A process describes truth as movement in and through substance, rather than substances as fixed concepts or things. Since Whitehead's work, process is distinguished as describing entities that arise or coalesce in becoming rather than being simply dialectically determined from prior posited determinates. These entities are referred to as complexes of occasions of experience. It is also distinguished by not necessarily being oppositional in operation. A process may be integrative, destructive or both together, allowing for aspects of interdependence, influence and confluence and addressing coherence in universal as well as particular developments. Additionally, instances of determinate occasions of experience, while always ephemeral, are nonetheless seen as important to define the type and continuity of those occasions of experience that flow from or relate to them.

Chapter 4
Epistemology

Epistemology is the theory of knowledge and metaphysics. However, philosophers disagree about what is knowledge, about how one obtains it and even about whether there is any to be gotten. The theory of reality has traditionally competed for the primary role in philosophical inquiry. Sometimes epistemology has won and sometimes metaphysics, depending on the methodological and substantive presuppositions of the philosopher.

The discussion of epistemology began with Plato's *Theaetetus,* if not earlier. Descartes' *Meditations* formulated the problems, which subsequent epistemologists have taken seriously, and philosophers have worked on the agenda that Descartes instigated. A great deal of the energy behind traditional epistemological inquiries was generated by the thought that perhaps human knowledge has no foundation. Lacking a foundation, we are reduced, so it was believed, to scepticism or, even worse, solipsism (Landesman 1997).

Thus, epistemology is the study of the nature and scope of knowledge and justified belief. It analyses the nature of knowledge and how it relates to similar notions, such as truth, belief and justification. It also deals with the means of creation of knowledge as well as scepticism about different knowledge claims. It is essentially about issues concerning the creation and dissemination of knowledge in particular areas of inquiry. Much debate in epistemology centres on four areas: (1) the philosophical analysis of the nature of knowledge and how it relates to such concepts as truth, belief and justification (Steup 2017), (2) various problems of scepticism, (3) the sources and scopes of knowledge and justified belief and (4) the criteria for knowledge and justification. Epistemology addresses such questions as what makes justified beliefs justified (Steup 2017), what does it mean to say that we know something and fundamentally how do we know that we know.

Therefore, the epistemologist asks what we know; the metaphysician asks what is real. Some philosophers have begun with an account of the nature of reality and then appended the theory of knowledge to account for how people know that reality. Plato, for example, reached the metaphysical conclusion that abstract entities, or

K. U. Koskinen, R. Breite, *Uninterrupted Knowledge Creation*, SpringerBriefs in Business, https://doi.org/10.1007/978-3-030-57303-4_4

forms, such as triangular or justice, are real and all else is mere appearance. Plato also held that the real is knowable, and he inquired into how people might know this reality. Aristotle, on the contrary, held that individual substances, such as individual statues or animals, are real and inquired as to how people might have knowledge, especially general knowledge, concerning these substances. In the opinion of Lehrer (1990), it is hardly surprising that Plato and Aristotle produced vastly different theories of knowledge when they conceived of the objects of knowledge in such different ways. People might refer to their common approach, starting with metaphysics, as metaphysical epistemology. The problem with this approach is that the metaphysical epistemologist uncritically assumes that people know the reality posited and only concern themselves with what such knowledge is like.

Descartes sought certainty, as did many of his successors. The basic propositions that were to constitute the ultimate foundations of knowledge were thought to be arrived at by processes deemed to be infallible or immune to error. However, with a few exceptions, the contents of individual belief systems lack the absolute certainty that Descartes thought they required if knowledge were to be possible.

According to Landesman (1997), the repudiation of the quest for certainty does not necessarily undermine the quest for the foundation of knowledge. Even if there are no, or only a very few, absolutely secure basic propositions, it is not necessary for a proposition to be basic for it to be secure. Instead of supposing that basic propositions are certain and that they lend their certainty to those derived from them, we can suppose instead that they possess a degree of credibility and that they transmit to derived propositions whatever degree of credibility they possess. People can search for the foundations in good conscience while dropping the quest for certainty.

To sum up, epistemology concerns the nature of our knowledge about the world. How is knowledge different from mere opinion? What are the criteria of validity for knowledge? Can knowledge be objective, and what does that mean? What are the limits of knowledge? According to Maturana (1974), people can have no access to an independent world with which to compare our theories and descriptions. How, then, are we to judge their truth or validity? The study of philosophical epistemology is the study of efforts to gain understanding or knowledge of the nature and scope of human knowledge.

4.1 Epistemological Assumptions

Differences in epistemology are manifested by different ways to categorize knowledge. This means, for example, that, by uncovering the epistemological roots of a system, one can better understand the characteristics of knowledge creation needed in that system. 'In order to manage knowledge assets, we need not merely to identify them but to understand them—in depth—in all their complexity: where they exist, how they grow, how individuals' actions affect their viability' (Leonard-Barton 1995, p. xii). According to Venzin et al. (1998), being familiar with different possible epistemologies means having a better understanding of the limitations of each approach. The following three subsections provide short illustrations of the cognitivist, connectionist and autopoietic epistemologies (cf. Varela et al. 1991 von Krogh and Roos 1995a).

4.1.1 Cognitivist Epistemology

The traditional cognitivist epistemology is based on the idea that the human mind has the ability to represent exactly the reality in a way that corresponds to the outer world, be it objects, events or states. This is also frequently referred to as the intentionally of the mind (cf. Goldman 1986). Broadening the idea, systems are considered to be entities that produce knowledge by formulating increasingly accurate representations of their predefined worlds. Because knowledge is seen as a representation of these worlds, knowledge accumulation and dissemination are the major knowledge creation activities in a system. The more knowledge a system can gather, the closer the presentation is to reality.

Learning in the cognitivist epistemology means improving the representations of the world by assimilating new experiences (Varela 1979; von Krogh et al. 1996). According to Bruner and Anglin (1973, p. 397), an individual actively constructs knowledge by relating incoming information to a previously acquired frame of reference. In other words, when gathering information from the external environment, an individual stores facts, relates them to existing experiences and creates a picture of the world. The world is considered to be a pre-given object, event or state, which can be perceived in an objective way. What varies from one individual to another is the ability to represent reality. The truth of knowledge is understood as the degree to which an individual's inner representation corresponds to the world outside. As new things are learned, this truth will constantly be improved.

4.1.2 Connectionist Epistemology

Representationism, as it has been described in cognitivist epistemology, is still prevalent in connectionist epistemology (von Krogh and Roos 1995a). In connectionism, however, the rules on how to process information are not universal but vary locally. Systems are seen as self-organized networks that are composed of relationships and driven by communication (Varela et al. 1991; Mingers 1995). The main method in connectionist epistemology is to look at relationships and not to focus on the individual or the entire system. The connectionist's models are built on a large number of integrating units that are able to influence one another by sending activation signals down interconnecting pathways. Systems are seen as networks. Like cognitivists, connectionists consider information processing to be the basic activity of the system. Connectionists see the process of shaping a system as dependent not only on the stimuli entering the system but also on the system itself. Relationships and communication are the most important issues of cognition.

The cognitivist and connectionist epistemologies share two assumptions. First, an individual or a system is directed to resolving a task. This means that an individual or a system must identify and represent that task as an inner creation of the cognitive system. Second, information processing is the basic activity of an individual or a system. For an individual, information is taken in from the environment through senses and will activate various components in the network of components that compose the individual (von Krogh and Roos 1995a).

However, the cognitivist and connectionist epistemologies also differ. While cognitivists assume that information processing depends only on stimuli from the environment, connectionists claim that it may also arise from within the system itself. The two epistemologies also assume that systems acquire representations in different ways. Cognitivists regard learning as a process of creating increasingly accurate representations of the external world. Connectionists understand representation as resulting from global states in a history-dependent system (von Krogh and Roos 1995a). The network as a whole learns from perceived patterns in its environment (Mingers 1995; Maula 2006).

4.1.3 Autopoietic Epistemology

Maturana and Varela are always aware of the epistemological implications of their ideas. They strongly maintain the distinction between the actual operational domain of an organism and the domain of descriptions of an observer. 'Everything said is said by an observer to another observer, who can be himself' (Maturana 1975, p. 6).

Compared with cognitivist and/or connectionist epistemology, autopoietic epistemology provides a fundamentally different understanding of the input coming from outside a system (e.g. Hall 2005). Input is regarded not as knowledge but as data; that is, knowledge is data but in a certain context. This means that knowledge cannot be transferred directly from an individual to another individual, because data have to be interpreted by the receiving individual before they become knowledge. According to autopoietic epistemology, information does not equal knowledge, but it is a process that enables knowledge creation and sharing to take place. Von Foerster (1984, p. 193) states that '... information is the process by which knowledge is acquired'. That is, books—for example, this book—manuals, memos, computer programs, and so on are data, not information.

As stated above, the autopoietic system is self-referential rather than having an input–output relationship with the environment. This means that its knowledge structure is made up of closed components of interactions that make reference only to them; that is, in this sense, the autopoietic system is autonomous. However, although the autopoietic system is autonomous, it will be perturbed by changes in its environment. For example, when an individual interacts in a recurrent manner, data creation elsewhere reaches him or her as perturbations. These perturbations trigger information processes in that individual (i.e. in the receiving system). This means that the perturbations trigger learning but do not specify it. The individual's own knowledge structure (i.e. cognitive map) determines which perturbations are allowed to enter the system and what changes in the existing knowledge structure are available at a given point in time.

For example, when a teacher delivers a speech, two students build different knowledge. The transmission by the teacher is the same for both, but the knowledge created is different. Knowledge therefore cannot be transmitted but only created or produced with the help of existing knowledge (Vicari and Troilo 1999). That is, the only way to acquire new knowledge (i.e. to learn) is to utilize existing knowledge.

4.2 Summary

The field of system studies has not paid much attention to the fundamental issues of epistemology. Knowledge has mostly been taken for granted, often as a fuzzy and substitutable concept. Therefore, this chapter has described three different epistemologies, namely the cognitivist, connectionist and autopoietic epistemologies. Some of the key factors are the following:

- The traditional cognitivist epistemology is based on the idea that the human mind has the ability to represent the reality in exactly the way that corresponds to the outer world.
- The connectionist epistemology is based on a large number of integrating units that are able to influence one another by sending activation signals down interconnecting pathways. Therefore, systems are seen as networks. Like cognitivists, connectionists consider information processing to be the basic activity of the system.
- Unlike the cognitivist or the connectionist epistemology, the autopoietic epistemology does not claim that the world is a pre-given but that cognition is a creative function. Thus, knowledge is a result of the autopoiesis, that is, the self-production process.
- In this book, the autopoietic epistemology is the basis of the understanding of knowledge production. The choice is based on the idea of presenting a fresh and alternative observational scheme for the understanding of knowledge creation in systems.

All in all, in the opinion of Goldman (1986), the central question asked in epistemology is how individuals or social entities know, or, in other words, through which processes individuals or social entities come to know of, the world. However, this is not enough. No epistemological investigation could leave the question of the nature of knowledge unanswered. Thus, epistemology analyses the nature of knowledge and how it relates to similar notions such as truth, belief and justification. It also deals with the means of creation of knowledge as well as scepticism about different knowledge claims. It is essentially about issues concerning the creation and dissemination of knowledge in particular areas of inquiry.

Chapter 5
Knowledge

The recent interest in the role of knowledge in working life and the discussion that this has triggered have largely avoided the epistemological question; that is, what is knowledge? Even the pragmatic question concerning the meaning that people in the West attach to the concept has been avoided. On the other hand, many dichotomies have been suggested. Most commonly, a distinction is made between practical and theoretical knowledge. One variant distinguishes between experiential knowledge and reported knowledge. Another refers to intimate knowledge as opposed to declared knowledge. Tacit knowledge as opposed to codified knowledge is another dichotomy, which often appears in discourse. Distinctions of this kind are useful for practical purposes, but they are also risky, since they tempt people to make radical simplifications of a multi-faceted reality (Wikström et al. 1994).

Hence, knowledge is not the homogeneous or clearly defined concept that the current debate on its role in society might seem to imply. Even the idea of knowledge in creation allows several interpretations. Knowledge is conceived both as an objective and as a subjective phenomenon—that is, it embraces facts and principles that exist independently of the consciousness of the individual—but awareness and recognition of such facts are also an aspect of knowledge. Whether facts and principles actually exist as part of an objective reality or whether they are constructs of the human mind is essentially a philosophical and epistemological question. The predominant trend in modern Western thought is towards the constructivist interpretation.

5.1 Differences Between Data, Information and Knowledge

In everyday language, we use knowledge all the time. Sometimes we mean know-how, while other times we are talking about wisdom. On many occasions, we even use it to refer to information. Part of the difficulty of defining knowledge arises from

K. U. Koskinen, R. Breite, *Uninterrupted Knowledge Creation*, SpringerBriefs in Business, https://doi.org/10.1007/978-3-030-57303-4_5

its relationship with two other concepts, namely data and information. These two terms are often regarded as lower denominations of knowledge, but the exact relationship varies greatly from one example to another.

Data are the **facts of the world.** For example, one has fair hair whether this is written down somewhere or not. All of this constitutes data. In many ways, data can be thought of as a **description of the world.** We can perceive these data with our senses, and then the brain can process them.

Furthermore, human beings have used data for as long as they have existed to form knowledge of the world. Until they started using information, all they could use was data directly. If one wanted to know how tall one was, one would have to look at oneself. Our knowledge was limited by our direct experiences.

Information allows us to expand our knowledge beyond the range of our senses. We can **capture data in information** then move it about so that other people can access it at different times. If I take a picture of someone, **the photograph is information, but what he or she looks like is data.** I can move a photo of someone around by sending it to other people via e-mail and so on. However, I am not actually moving *the person* around—or *what the person looks like.* I am simply allowing other people, who cannot see the individual directly from where they are, to know what the person looks like. If I lose or destroy the photo, this does not change how the person looks.

Knowledge is what we know. Think of this as a map of the world that we build inside our brain. Like a physical map, it helps us to know *where* things are—but it contains more than that. It also contains our beliefs and expectations. If I do this, I will probably get that. Crucially, the brain links all these things together into a giant network of ideas, memories, predictions, beliefs and so on. It is from on this map that we base our decisions, **not the real world itself**. Our brain constantly updates this map from the signals coming through our eyes, ears, nose, mouth and skin.

One **cannot currently store knowledge** in anything other than a brain, because a brain connects it all together. Everything is inter-connected in the brain. Computers are not artificial brains. They do not **understand** what they are processing, and they cannot make **independent decisions** based on what one tells them.

Thus, generally people talk about knowledge in the subjective sense, that is, in terms of an individual and what he or she knows. However, knowledge also exists in an objective sense, as embodied in books, systemic practices and procedures, the Internet and so on, but this ultimately depends on the knowledge of the individuals and groups who generate it and then access it. Polanyi (1958) argues that all knowledge has a personal dimension: that all knowing is personal knowing (Mingers 2010).

Knowledge is always changing. For example, in a system, like a business firm, knowledge changes around products, services, processes, technology, structures, roles and relationships. No sooner do people think they have identified a pattern of knowledge than a new one seems to appear. Then, if the basic components of systemic intelligence are always changing, how do people organize to support this?

Anyway, all agree that knowledge is valuable, but the agreement about knowledge tends to end there (Lehrer 1990). Philosophers disagree about what knowledge is, about how one obtains it and even about whether there is any to be gained. In the opinion of Haraway (1991), knowledge is situational, which means that it is always qualified as knowledge from a particular point of view. Knowledge is situational because it is inherently social in nature. It serves to establish relations in society. Therefore, it is never value neutral but always already emergent from specific social interests and concerns (e.g. Lam 1997; Sole and Edmondson 2002). More specifically, scientific knowledge is forcefully rejected by the analysis of techno-scientists (Haraway 1991). Gherardi and Nicolini (2001, p. 44) write: 'Every attempt to label something as knowledge is made by a specific social community belonging to a network of power relations, and not by a world consisting purely of ideas. Hence, no knowledge is universal or supreme. Instead, all knowledge is produced within social, historical, and linguistic relations grounded in specific forms of conflict and the division of labour.'

Then, the course of erotetics in relation to questions' change is no less dramatic than that of cognitive change in relation to knowledge. A change of mind about the appropriate answer to a question will unravel the entire fabric of questions that presupposed this earlier answer. If we change our mind regarding the correct answer to one member of a chain of questions, then the whole of a subsequent course of questioning may well collapse.

Questions cluster together in groupings that constitute a line of inquiry. They stand arranged in duly organized and sequential families, the answering of a given question yielding the presuppositions for yet further questions that would not have arisen had the former questions not been answered.

In one sense, to know means to have some special form of competence. Thus, to fix up a car is to be competent to repair a car. If a person is said to know how to do something, it is this competence sense of knowing that is usually involved.

Another sense of knowing is that in which the word means being acquainted with something or someone. When someone says that he or she knows Adam, he or she means that he or she is acquainted with Adam. The sentence 'I know the village' is more difficult to disambiguate. It might mean simply that I am acquainted with the village and hence have the acquaintance sense of knowing or it might mean that I have the special form of competence needed to find my way around the village. This example illustrates the important fact that the senses of knowing that people are distinguishing are not exclusive. In other words, the term 'know' may be used in more than one of these senses (Russell 1910–1911).

The third sense of knowing is that in which to know means to recognize something as information. If I know that the stars are bound together by gravitational forces, then I recognize something, namely that the stars are bound by gravitational forces. It is often affirmed that to know something in the other senses of knowing entails knowledge in the information sense of knowing. Here we are concerned with knowledge in the information sense, because this sense is fundamental to human cognition and requires both for theoretical speculation and practical sagacity. To perform science, to engage in experimental inquiry and scientific ratiocination, one must be able to tell whether one has received correct information to obtain scientific knowledge of the world. This sort of knowledge extends beyond the mere possession of information. If you tell me something and I believe you, even though I have no idea whether you are a source of truth and correct information about the subject or a propagator of falsehood and deception, I may, if I am fortunate, acquire information when you happen to be informed and honest.

The cognitive opacity of real things means that we are not—and will never be—in a position to evade the contrast between things as we think them to be and things as they actually and truly are. Their susceptibility to ever more elaborate detail and to changes of mind regarding this further detail is built into our very conception of real things. To be real is to be something regarding which we can always, in principle, acquire further new information, which may not only supplement but even correct that which has previously been acquired. This view of the situation is bolstered rather than impeded once we abandon the cumulative view of knowledge acquisition for the view that new discoveries need not augment but can displace old ones. With further inquiry, we may come to recognize the error of our earlier ways of thinking about the things at issue. We realize that people will come to think differently about things from the way that we do—even when thoroughly familiar things are at issue—recognizing that scientific progress generally entails fundamental changes of mind about how things work in the world.

5.2 Processual Knowledge

To gain a better understanding of organizing in world on the move, it is necessary to understand everything as being in process. This implies that it is not tenable to assume any stable contexts as framing processes, since connecting is seen as the essence of organizing. In a world on the move, the ability to conceptualize the emergence of novelty as part of the process is important. Novelty cannot, however, be outside the process: it must be an integral part of the process, just like the recurrence of the known. Novelty implies connecting to untried possibilities, which may be distant from actors both in time and in space.

According to Hernes (2014), this is not, however, the way that processes have traditionally been considered in system studies. Traditional process views, such as that presented by Pettigrew (1997), seem to assume that processes are framed by relatively stable systemic contexts, which is what Hernes and Weik (2007) refer to as an exogenous view of processes. According to an exogenous view, flows such as actions, communication, behaviour and so on are assumed to be influenced by the external context of the process, which may consist of entities such as rules and natural conditions. At an analytical level, the view corresponds to seeing systems as consisting of processes within structures and seeing context as something within which interaction takes place.

Some researchers (e.g. Tsoukas 2005; Hernes 2008) suggest adopting a process perspective on knowledge, claiming that, to understand what knowledge is, we should focus our attention on the processes or practices of knowing. The process perspective emphasizes that knowledge is socially constructed; that is, knowledge is inherently social and embedded in practice.

According to Rescher (2000), human knowledge should be thought of as a process rather than merely a product. It is clearly not stable. Because ongoing inquiry leads to new and often dissonant findings and discoveries, knowledge emerges in phases and stages through processes that engender an ever-changing state of the art. The coordination between questions and bodies of knowledge means that, in the course of the cognitive progress, the state of questioning changes no less drastically than the state of knowledge. Cognitive change regarding answers inevitably carries in its wake erotetic change with respect to questions, since alteration in the membership of our body of knowledge will afford new presuppositions for further questions that were not available before.

Furthermore, according to Rescher (2000), inquiry is a dialectical process, a step-by-step exchange of query and response that produces sequences within which the answers to our questions ordinarily open up yet further questions. This leads to a erotetic cyclic process that determines a course of inquiry that is set by an initial, controlling question together with the ancillary questions to which it gives rise and the solutions to which are seen as facilitating its resolution.

Our knowledge of reality itself has a practical and thus processual dimension. One of the most significant and characteristic kinds of know-how is the knowledge of how to operate at the level of theory—how to conjure with theoretical knowledge over the range from obtaining it, to using it and to conveying it. Here we come to the whole domain of securing, recording, communicating and processing information. Achieving any sort of knowledge is itself one of the most extensive and significant forms of praxis in which our species is involved. That is, praxis is by its very nature something processual. Any adequate worldview must recognize that the ongoing progress of scientific inquiry is a process of conceptual innovation that always leaves various facts about the things of this world wholly outside the cognitive range of the inquirers of any particular period. Ideas cannot be grasped before their time—before the developmental unfolding of the beliefs and notions that alone can provide an intellectual entry into their domain. The novelty that arises with the emergence of new cognitive processes is crucial both to the nature and to the availability of our ideas. This dynamics of ideas strikingly marks the processual aspect of epistemology. Throughout the cognitive enterprise, we are confronted with an ever-changing state of the art. Knowledge is not a thing, let alone a commodity of a fixed and stable make-up. It is irremediably processual in nature, affected as deeply by the fluid nature of reality as anything else.

5.3 Epistemological Knowledge

Epistemic (knowledge) change over time relates not only to what is known but also to what can be asked. The accession of new knowledge opens up new questions, and, when the epistemic status of a presupposition changes from acceptance to abandonment or rejection, we witness the disappearance of various old ones through dissolution. A question solved in one era could well not even have been posed in another.

Epistemology—the theory of knowledge—and—metaphysics—the theory of reality—have competed for the primary role in philosophical inquiry. Sometimes epistemology has won and sometimes metaphysics, depending on the methodological presuppositions of the philosopher.

Some have denied that individuals know what is true or what is false, and they have remained sceptics. Scepticism will have a hearing, but here it is pursued by critical epistemologists. That is, it is assumed that people have knowledge, but what sort of knowledge do they have, and what is knowledge anyway? There are many sorts of knowledge, but only one, the knowledge that something is true, will be the concern here.

In view of the cognitive opacity of the real, we always do well to refrain from pretending to a cognitive monopoly or cognitive finality. This recognition of incomplete information is inherent in the very nature of our conception of a real thing. It is a crucial facet of our epistemic stance towards the real world to recognize that every bit of it has features lying beyond our present cognitive reach—in any present whatsoever.

The cognitivist and connectionist epistemologies assume that the world is pre-given and that the task of the cognitive system is to represent this world as correctly as possible. The cognitive system construes the world around it. The metaphor is that of a human mind mirroring nature (Rorty 1980). The cognizing system subsequently acts on these representations. According to Varela et al. (1991 p. 136), 'The ultimate court of appeal for judging the validity of (these) representation(s)' is the world in which we act, unsuccessfully or successfully.

Autopoiesis theory at a very general level suggests an alternative conception of the relationship between human knowledge and the world. The discussion that follows points to various properties of knowledge of human beings: knowledge is embodied, it is self-referential and allows for distinction making in observations and it is brought forth in a systemic setting.

5.4 Individual Knowledge

Human beings rely on their experiences and creativity in defining a problem and the possible solutions to the problem. A theory of knowledge rooted in autopoiesis theory suggests that knowledge is not abstract but embodied: everything known is known by somebody (Maturana and Varela 1980). As human beings confront new situations, experiences are gained through thinking, sensing and moving (von Krogh and Roos 1995a). Knowledge is formed through actions, perceptions and sensory processes (Merley-Ponty 1963; Schutz 1970; Varela et al. 1991). Autopoiesis theory also recognizes that human beings use past experiences to orient themselves in new situations. Thus, previous experiences will affect the new experiences gained.

Thus, the knowledge of human beings is embodied, self-referential, allows for distinction making in observations and is brought forth in a systemic setting (von Krogh and Roos 1995a). This view of embodied knowledge leads to a startling view of the relationship between the world and the knowledge of human beings. A key claim is that a situation, or world, and knowledge are structurally coupled and hence co-evolve. Knowledge enables people to perceive, act and move in a world, and, as they act, perceive and move, the world comes forth as a result of their actions and observations. In the words of Maturana and Varela (1980), knowledge is what brings forth a world. Furthermore, in the words of Schutz (1970), the world refers to subjective experience and comprehension. It is a world of somebody, namely the concretely experiencing individual.

The view of embodied knowledge also maintains the concern with autonomy that is critical for autopoiesis theory. Knowledge develops in an autonomous manner for human beings and thus cannot be transferred directly to other human beings. In other words, a human being's history is unique and structurally coupled with the world. Each human has his or her own history of movement and observation, his or her own pattern of structural coupled interaction with the world. As a result, the evolving knowledge, because it is formed in structural coupling, is also unique (von Krogh and Roos 1995a). This means that an individual's knowledge is a result of directly experiencing tasks through a history of structural coupling.

The concept of self-reference has strong implications for the way in which human knowledge is viewed. Knowledge is intimately connected to creativity, action, observation, hearing, smelling and so on. The broad repertoire of human activity contributes to knowledge. Cognitive processes refer to themselves. All knowledge will always be self-knowledge: when an individual knows (brings forth a world), this will reveal something about himself or herself (Morgan 1996). Even when individuals are acting spontaneously (Schutz), in hindsight their actions reveal something about themselves to themselves.

Therefore, in the case of an individual, '... knowledge is the individual ability to draw distinctions within a collective domain of action, based on an appreciation of context or theory, or both' (Tsoukas and Vladimirou 2001, p. 979). This means that an individual's capacity to exercise judgement is based on an appreciation of context in the ethno-methodological sense, that a social being is knowledgeable in accomplishing a routine and taken-for-granted task within a particular context as a result of having been through processes of socialization.

5.5 Systemic Knowledge

Systemic knowledge is all the knowledge resources within a system that can be realistically tapped by that system. It can therefore reside in individuals and groups or exist at the systemic level. This means that systemic knowledge resides in both the individual systemic members and in the relations among systemic members, that is, at the social level. In keeping with the anti-representational stand, at our disposal we find strong theoretical foundations with respect to individual human cognition and individualized knowledge. However, the definition of systemic knowledge receives very little consensus within the literature. Variations include the extent to which the knowledge is spread within the system as well as the actual make-up of this knowledge. Hatch (2010) defines it as follows: 'When group knowledge from several subunits or groups is combined and used to create new knowledge, the resulting tacit and explicit knowledge can be called organizational (i.e. systemic) knowledge'. Others present a broader perspective: 'individual knowledge, shared knowledge, and objectified knowledge are different aspects or views of organizational (i.e. systemic) knowledge' (Ekinge and Lennartsson 2000).

Furthermore, in the opinion of Tsoukas (2005), although most people intuitively identify knowledge with individual knowledge, it is not quite evident how knowledge becomes an individual possession and how it is related to individual action, nor it is clear in what sense knowledge merits the adjective systemic.

Thus, knowledge is the individual capability to draw distinctions, within a domain of action, based on an appreciation of context or theory or both. According to Tsoukas (2005), systems are three things at once: concrete settings within which individual action takes place; sets of abstract rules in the form of propositional statements; and historical communities. Systemic knowledge is the capability that the members of a system have developed to draw distinctions in the process of carrying out their work, in particular concrete contexts, by enacting sets of generalizations, the application of which depends on historically evolved collective understandings and experiences. The more propositional statements and collective understandings become instrumentalized (in Polanyi's sense of the term), the more new experiences are reflectively processed (both individually and collectively) and then gradually driven into subsidiary awareness, the more systemic members dwell in all of them and the more able they become to concentrate on new experiences on the operational plane.

5.6 Tacit Knowledge

Tacit knowledge is knowledge that people can make use of but that they cannot formulate and provide with proper expression (Polanyi 1958). According to Polanyi (1958), tacit knowledge is an effect of what Chomsky (1968) calls speech performance, that is, the ability to formulate experiences and ideas. As one may easily realize, speech performance capabilities are unevenly distributed across the population. Some people may be able to give very detailed and adequate accounts of their experiences, while others may be incapable of saying much about what they experience.

Indeed, tacit knowledge is context dependent and situation sensitive (Varela et al. 1991; Lyles and Schwenk 1992; Starbuck 1992; Koskinen et al. 2003). In the words of von Krogh et al. (1996, p. 164), '... knowledge depends very much on the point of observation. Where you stand or what you know determines what you see or what you choose to be relevant.' That is, on the basis of these authors, it is possible to conclude that tacit knowledge is not abstract but embodied in the individual's worldview. Rosenberg's (1982, p. 43) description of traditional technological knowledge, accumulated in crude empirical ways with no reliance on science, provides a good definition of tacit knowledge in technology companies: '... the knowledge of techniques, methods and designs that work in certain ways and with certain consequences, even when one cannot explain exactly why'.

Thus, tacit knowledge resides within the individual and is known but extremely difficult or in some cases impossible to articulate or communicate adequately (e.g. Newell et al. 2002). Polanyi (1958) suggests, for example, that we know more than we can articulate. Tacit knowledge is often referred to as know-how. It resides in our heads and in our practical skills and actions.

According to Polanyi (1966), we can recognize a face from among a thousand almost instantly, but we usually cannot explain how we recognize a face that we know. Tacit knowledge permeates our personal and work lives, enabling us to drive a car, enjoy a film or deal with a problem situation. In all such cases of personal knowing, '... the aim of a skilful performance is achieved by the observance of a set of rules which are not known as such to the person following them' (Polanyi 1958, p. 49): particular rules or elements, which partly exist as an emergent quality of knowing something as a whole. In the words of Polanyi (1966, p. 20), 'The skill of a driver cannot be replaced by a thorough schooling in the theory of the motorcar; the knowledge I have of my own body differs altogether from the knowledge of its physiology; and the rules of rhyming and prosody do not tell me what a poem told me, without any knowledge of its rules'.

Further, tacit knowledge may be likened to knowing that is in our action, '... implicit in our patterns of actions, and in our feel for the stuff with which we are dealing' (Schön 1983, p. 54). Schön defines this knowing in action as having the following properties:

- There are actions, recognitions and judgements, which we know to carry out spontaneously; we do not have to think about them prior to or during their performance
- We are often unaware of having learned to do these things; we simply find ourselves doing them.

In some cases, we were once aware of the understandings, which were subsequently internalized in our feeling for the stuff of action. In other cases, we may never have been aware of them. In both cases, however, we are usually unable to describe the knowing that our action reveals. In the opinion of Leonard-Barton and Sensiper (1998), there are three main ways in which tacit knowledge can potentially be exercised to the benefit of the system:

- Problem solving. The most common application of tacit knowledge is for problem solving. The reason that experts on a given subject can solve a problem more readily than novices is that the experts have in mind a pattern born of experience, which they can overlay on a particular problem and use to detect a solution

quickly. The expert recognizes not only the situation in which he or she finds himself or herself but also the action that might be appropriate for dealing with that situation. Writers on the topic note that intuition may be viewed most usefully as a form of unconscious pattern-matching cognition.
- Problem finding. The second application of tacit knowledge is to the framing of problems. Some researchers distinguish between problem finding and problem solving. Problem solving is linked to a relatively clearly formulated problem within an accepted paradigm. Problem finding, on the other hand, tends to confront the person with a general sense of intellectual unease, leading to a search for better ways of defining or framing the problem. Creative problem framing allows the rejection of the obvious or usual answers to a problem in favour of asking a wholly different question. Intuitive discovery is often not simply an answer to the specific problem but an insight into the real nature of the dilemma.
- Prediction and anticipation. The deep study of a subject seems to provide an understanding, only partially conscious, of how something works, allowing an individual to anticipate and predict occurrences that are subsequently explored very consciously. Histories of important scientific discoveries highlight that these kinds of anticipations and reliance on inexplicable mental processes can be very important in invention. Authors writing about the stages of creative thought often refer to the preparation and incubation that precede flashes of insight.

While tacit knowledge is a personal resource, researchers (e.g. Wenger 1991; Brown and Gray 1995) find that teams as well as whole systems can be usefully thought of as possessing knowledge that has the characteristics of tacit knowledge, that is, knowledge that is hard to document but is dispersed among multiple actors who interact with each other and with the physical, cultural and social dimensions of their task and system setting. Studies such as those of Orr (1990) and Brown and Duguid (1991) challenge the traditional assumptions that learning and knowing implies individual mastery and that everything that is knowable can be made explicit.

Instead of treating knowledge as being explicit and individually acquired, knowledge in systems is often tacitly shared by members of social groups. In the words of Brown and Gray (1995, p. 80), 'With individuals, tacit knowledge means intuition, judgement, common sense—the capacity to do something without necessarily being able to explain it. With groups, tacit knowledge exists in the distinct practices and relationships that emerge from working together over time—the social fabric that connects communities of knowledge workers.'

According to Wenger (1991), a group holds tacit knowledge as a community that forms around a shared practice. Members of such communities of practice participate in a shared practice informally but legitimately. The community of practice provides a context in which the meaning of objects, problems, events and artefacts is constructed and negotiated and in which people live, work, communicate and understand the environment and themselves. Communities of practice emerge naturally from the system's web of interactions and need not be formally controlled or designed. By reconceiving systems as comprising communities of practice, working, learning and innovation are integrated in a unified view (Brown and Duguid 1991).

All in all, tacit knowledge represents knowledge—and meaning—based on the experience of individuals. It is expressed in human actions in the form of evaluations, attitudes, points of view, commitments, motivation and so on (e.g. Polanyi 1966; Nonaka and Takeuchi 1995; von Krogh and Roos 1995a; Koskinen 2000). Usually it is difficult to express tacit knowledge directly in words, and often the only ways of presenting it are through metaphors (e.g. Tsoukas 1991), drawings and different methods of expression that do not require the formal use of language.

5.7 Explicit Knowledge

Explicit knowledge, unlike tacit knowledge, can be embodied in a code or a language, and as a consequence it can be communicated easily. In other words, the meanings representing explicit knowledge in the worldview are rather clear and conscious, and therefore an individual can easily retrieve them from his or her worldview. They represent knowledge in a narrow sense. The code may be words, numbers or symbols, like grammatical statements, mathematical expressions, specifications, manuals and so forth (Nonaka and Takeuchi 1995). For example, explicit knowledge implies factual statements about such matters as material properties, technical information and tool characteristics.

However, there is no dichotomy between tacit and explicit knowledge, but tacit and explicit knowledge are mutually constituted (Tsoukas 1996). In other words, they should not be viewed as two separate types of knowledge, but these kinds of meanings are intermingled in the worldview. This means that, for any explicit knowledge, there is some tacit knowledge. That is, explicit knowledge is an extension of tacit knowledge to a new level (Mooradian 2005). Hence, if there is value in identifying tacit knowledge, it is in relation to making explicit knowledge understandable.

5.8 Sensemaking

Sensemaking is defined in various ways by different researchers. March and Olsen (1976, p. 56) see sensemaking as part of experiential learning in which '... individuals and systems make sense of their experience and modify behaviour in terms of their interpretations'. Starbuck and Milliken (1988, p. 51) observe that '... sensemaking has many distinct aspects—comprehending, understanding, explaining, attributing, extrapolating, and predicting, at least ... what is common to these processes is that they involve placing stimuli into frameworks (or schemata) that make sense of the stimuli'. According to Choo (1988), people in systems are continuously trying to understand what is happening around them. They first have to make sense of what is happening in their environments to develop a shared interpretation that can serve as a guide to action. Thus, sensemaking is often thought of as belonging to a larger process of systemic adaptation that also includes scanning the environment, interpreting and developing responses.

Thus, sensemaking is the process by which people give meaning to their collective experiences. It is defined as 'the ongoing retrospective development of plausible images that rationalize what people are doing' (Weick et al. 2005, p. 409). The concept was introduced to organizational (i.e. system) studies by Karl E. Weick in the 1970s and has affected both theory and practice. Weick intended to encourage a shift away from the traditional focus of system theorists on decision making and towards the processes that constitute the meaning of the decisions that are enacted in behaviour.

Furthermore, sensemaking is the ability or an attempt to make sense of an ambiguous situation. More exactly, sensemaking is the process of creating situational awareness and understanding in situations of high complexity or uncertainty to make decisions. It is a motivated, continuous effort to understand connections (which can be among people, places and events) to anticipate their trajectories and act effectively.

Sims and Gioia (1986) argue that the heaviest and most dominant work in systems is sensemaking (Weick 1995) rather than decision making. Whenever possible, before taking a decision, one needs to give sense to and make sense of the world. As a matter of fact, situations in which problems and actions are well structured and defined, regardless of their solvability, are the exception and not the rule. Rather paradoxically, the emphasis in sensemaking is sustained even by supporters of autopoiesis (Luhmann 1990; von Krogh and Roos 1995b; Maturana 1978; Winograd and Flores 1986), but it is inexplicable how it corresponds to invariance.

Weick (1979) encapsulates the main sensemaking recipe in the question 'How can one know what he or she thinks until he or she sees what he or she says?' The recipe suggests that people in systems are continually engaged in talk to find out what they are thinking and to construct interpretations of what they are doing. The recipe is executed in connected sequences of enactment–selection–retention.

Enactment is the process by which individuals in a system actively create the environments that they face and to which they then attend. The enactment process begins as a result of noticing a change or discrepancy in the flow of experience. Raw data about these environmental changes form the input for the process. Individuals isolate some of these changes for closer attention by bracketing and labelling portions of the experience or by taking some action to create features of the environment to which to attend. In this way, '... managers (i.e. individuals) construct, rearrange single out, and demolish many objective features of their surroundings ... people, often alone, actively put things out there that they then perceive and negotiate about perceiving. It is that initial implanting of reality that is perceived by the word enactment' (Weick 1979, pp. 164–165). The output of enactment is a set of equivocal, uninterpreted raw data, which supply the base material for the other sensemaking processes.

Selection is the process by which the people in a system generate answers to the question 'What is going on here?' (Weick 1979). What the selection process chooses are the meanings that can be imposed on the equivocal data from the enactment process. Possible meanings come from meanings and interpretations that have proven to be sensible in the past, as well as from '... patterns implicit in the enactments themselves' (Weick 1979, p. 175). Past interpretations are used as templates that are laid over current data to reveal plausible configurations. Selection, based on an assessment of the degree of fit, is necessary, because many of the possible meanings would be inapplicable to or inconsistent with the current data. The result of the selection process is an enacted environment that is meaningful in that it provides a cause-and-effect explanation of what is taking place.

Retention is the process by which the products of successful sensemaking, that is, enacted or meaningful environments, are stored so that they may be retrieved on future occasions as possible meanings to be imposed on new equivocal situations. Retained meanings are stored as enacted environments that are '... a punctuated and connected summary of a previously equivocal display' (Weick 1979, p. 131) or as cause maps that identify and label variables and connect the variables in causal relationships (Weick 1979, p. 132).

The sensemaking model sees the system as trying to make sense of its equivocal environment. Members look back on their actions and experiences and enact or construct their own perceptions of the environment. Sensemaking is retrospective in that members can only interpret what they have already done or what has happened. The outputs of sensemaking are enacted environments or shared interpretations that guide action.

An important corollary of sensemaking is that individuals/systems behave as interpretation systems. What is being interpreted is the individuals'/systems' external environment, and how the individual/system approaches the interpretation depends on how analysable the environment is perceived to be and how actively the individual/system intrudes into the environment to understand it. Equability is reduced by the participants, who extensively discuss ambiguous data cues and so arrive at a common interpretation of the external environment.

Thus, making sense, or constructing meaning from what has been sensed about the environment, is problematic, because the data about the environment are ambivalent and therefore subject to multiple interpretations. Selecting an appropriate interpretation is hard, because each individual sees different parts of the environment as interesting, depending on the individual's values, history and experience. Whereas sensing is gathering sufficient data to reduce environmental uncertainty, sensemaking involves choosing and agreeing on a set of meanings or interpretations to reduce the ambiguity in environmental cues. Unlike scanning, which can be designed as a systemic and structural activity, sensemaking is inherently a fluid, open, disorderly, social process. The basic mode of sensemaking is discourse, for it is through talk that systemic members find out what all others think, and it is through talk that people persuade, negotiate and reshape their points of view. Sensemaking is further complicated by the possibility that the system can or wishes to intrude actively into the environment to produce, influence or modify parts of it.

As a summary, one may say that sensemaking is a continuous social process in which individuals look at elapsed events, bracket packets of experience and select particular points of reference to weave webs of meaning. The result of sensemaking is enacted in a meaningful environment, which is a reasonable and socially credible rendering of what is taking place. The central problem in sensemaking is how to reduce or resolve ambiguity and how to develop shared meaning so that the system may act collectively.

5.9 Meaning

As originally developed by Husserl (1948, 1950), the concept of meaning denotes the surplus of references to the other possibilities of an experience or action. The meaning of a knife, for example, is its reference to actions and experiences like cutting, stabbing, eating, operating, cooking and so on. Thus, the knife is not only a knife as such but a knife with regard to something beyond the knife (Seidl 2005). In this context, Luhmann (1995a, p. 60) writes: 'Something stands in the focal point, at the centre of intention, and all else is indicated marginally as the horizon of an "and so forth" of experience and action'.

According to Seidl (2005), meaning is the difference between the real and the possible or between actuality and potentiality. A momentarily actual experience or action refers to other momentary, not actual but possible, experiences. The significance of this distinction becomes clear if one looks at it from a dynamic perspective. While the one side of the distinction indicates what is momentarily actual, the other side indicates what could consequently become actual (Luhmann 1995a, p. 74).

Thus, meaning is an event that disappears as soon as it appears. It marks a merely temporal point after which something else has to follow. The combination of this instability with the co-presentation of possible ensuing events results in the particular dynamic of meaning. Every meaning event disappears as soon as it takes place, but it produces further meaning events to succeed it. For Luhmann, this auto-agility of meaning events is 'autopoiesis par excellence'.

According to Pihlanto (2005), all the knowledge that an individual has acquired is accumulated into his or her worldview in the form of meanings. Meanings can be classified into different categories; therefore, knowledge can also be categorized accordingly. Knowledge can be defined in both a narrow and a wide sense. The former contains scientific research results and other more or less factual types of knowledge. In a wide sense, tacit knowledge can also be considered as knowledge.

For instance, intuition is a type of meaning and therefore knowledge in a wide sense. Further, such mental conditions as feeling, belief and will are meanings and therefore relevant to an individual's understanding of phenomena. In a wide sense, all types of meanings are knowledge, because an individual understands what the world is like on the basis of these types of meanings.

In addition to the complicated intermingling of different types of meanings, meanings are not always clear and unambiguous: in many cases, they may be unclear, ill-structured, distorted or even erroneous, but they are nevertheless meanings, on the basis of which a decision maker understands the issue at hand in one way or another. Meanings are not only concrete in content but may also be abstract or ideal (e.g. mathematical relationships), which means that the meaning has emerged not from any real object but instead from an abstract object. Moreover, in the mind, a continuous process of restructuring of meanings occurs, in which meanings are also often forgotten, fading into unconsciousness, possibly to be retrieved later.

To sum up, the experience of meaning is not a mechanical realization of a routine or a procedure. This means that people's engagement in practice may have patterns, but it is the production of such patterns anew that gives rise to an experience of meaning. All that people do and say may refer to what has been done and said in the past, yet they produce a new situation, an impression or an experience: they produce meanings that extend, redirect, dismiss, reinterpret, modify or confirm—that is, negotiate a new—the histories of meanings of which they are a part. In that sense, living is a constant process of negotiation of meaning (Wenger 1998).

5.10 Experience

Experience is the knowledge or mastery of an event or subject gained through involvement in or exposure to it. Terms in philosophy, such as empirical knowledge or a posteriori knowledge, are used to refer to knowledge based on experience. A person with considerable experience in a specific field can gain a reputation as an expert. The concept of experience generally refers to know-how or procedural knowledge rather than propositional knowledge: on-the-job training rather than book learning.

The notion of experience may refer, somewhat ambiguously, both to mentally unprocessed immediately perceived events and to the purported wisdom gained in subsequent reflection on those events or interpretation of them. Some wisdom–experience accumulates over a period of time, though one can also experience and gain general wisdom–experience from a single specific momentary event. One may also differentiate between (for example) physical and mental experiences.

Idea formation is a salient characteristic capacity of intelligent beings, and experience—that is, the interaction between minds and nature—is the pivotal mode of process here. The empiricists in this were right at any rate: all our ideas have a basis in experience. However, they were wrong in thinking that this experience has to be sensory rather than, more broadly speaking, intellectual. As Kant stressed, empiricists neglect the crucial fact that all our experience is conceptual in nature. To be sure, our experience is intentional, it is always of something and it is accordingly object oriented. However, the object at issue need not be something substantive. Processes as such also count: in fact, they are central.

Mental experience involves the aspect of intellect and consciousness experienced as combinations of thought, perception, memory, emotion, will, thought and imagination, including all unconscious cognitive processes. The term can refer, by implication, to a thought process. Mental experience and its relation to the physical brain form an area of philosophical debate: some identity theorists originally argued that the identity of the brain and mental states held only for a few sensations. Most theorists, however, generalized the view to cover all mental experiences.

According to Ushenko (1986), what we actually experience is not a matter of the sense data of earlier perception theorists, placed in a static co-presence like letters on a printed page. We experience the powers of things: not just the flames but their radiating warmth; not just the redness of Mike's face but his anger. We observe not just things but their activities and operations, and this brings their process into the orbit of experience. Then, human observation is a commerce with the world's processes, a dynamic encounter that is not just a matter of noting its fixed features.

5.11 Memory

A systemic memory tends to survive over time as a composite sediment from successive articulations that synthesize the very meaning of the system. Memory is strengthened by particular events and rituals and involves symbolic acts and artefacts. Memory, however, although it accumulates as a retrospective phenomenon, is applied prospectively in an ongoing present (Hernes 2014).

Thus, memory is central to both continuity and change. Bergson (1922), like Deleuze (2004), conceives of memory and past in inner experience as a pure ongoing process that is not subject to preservation in any artificial form. Bergson (1922, p. 5) makes the point as follows: 'And as the past grows without ceasing, so also there is no limit to its preservation. Memory, as we have tried to prove, is not a faculty of putting away recollections in a drawer, or of inscribing them in a register. There is no register, no drawer; there is not even, properly speaking, a faculty, for a faculty works intermittently, when it will or when it can, while piling up of the past upon the past goes on without relaxation.'

Most memory in systemic life is indirect in the sense that it refers to the experience of others, which, as Schutz (1967) writes, obliges actors to imagine the context of other actors' choices. Hence, evoking memory may take different forms, as shown by Schulz and Hernes (2013). Moreover, this process can be carried out more or less consciously. A useful distinction to draw is to view memory either as being unconsciously embedded in system actions, routines or programmes or as conscious attempts to evoke past experiences (Cohen 1991). The great achievement of systemic life lies in its ability to make it possible for multiple actors to perform coordinated and often complex tasks without thinking about them. Practical articulation is performed almost subconsciously and provides people with what they need to keep doing what they are doing. This is what is referred to by some writers as procedural memory. In systemic life, procedural memory is embedded in skills, practices, routines, or programmes. A key characteristic of procedural memory, as Moorman and Miner (1998) point out, is that it becomes automatic or unconsciously accessible. People may be aware that they perform practices that were conceived in the past but are ignorant of how and why those practices emerged. This is also why procedural memory refers to process memory or memory of underlying skills for performing tasks (Moorman and Miner 1997).

The counterpart to procedural memory is declarative memory (Cohen 1991; Moorman and Miner 1997; Singley and Anderson 1989), which may be found in facts, events, or propositions (Cohen 1991; Moorman and Miner 1998; Singley and Anderson 1989) that are evoked and thus may become the object of reinterpretation. Declarative memory refers to a conscious search into the past to understand the context of past experiences and requires inter-subjective articulation. Occasions of evoking declarative memory are likely to become decisive events as they both provide a mirroring of events in a way that has not previously been undertaken and by the same token become perceived as central to a changing event formation. Whereas procedural memory provides for the continuity of a certain culture, declarative memory may enable change, because it allows for reflection on the articulated practices that give rise to the prevailing culture. Change requires conscious efforts to evoke experiences to make way for the modification or addition of meaning structure elements.

5.12 Summary

Knowledge in different contexts is a concept that is extremely meaningful, promising and hard to pin down. This chapter has explored notions of knowledge, knowledge types and concepts that are based on knowledge in its large sense. Some of the claims and suggestions include the following.

Knowledge is an individual's perception, skills and experience, which are all dependent on the experiences that the individual's worldview contains in the form of meanings. This means that knowledge involves the individual combining his or her experience, skills, intuition, ideas, judgements, context, motivations and interpretations.

The traditional way to categorize knowledge is to make a distinction between data, information and knowledge. However, here we understand these terms by stressing the human dimension, that is, that data are raw knowledge, information is an interpretation process and knowledge is located in the worldview of an individual. Moreover, we put the emphasis on the categorization according to which knowledge is divided into individual, systemic, tacit, explicit, product, societal, expert and social knowledge.

The central insight of the knowledge in systems is that knowledge inputs are necessarily embedded in a context—cognitive and behavioural, individual and social—which powerfully constrains their discovery, their transfer from one set of actors to another and their usefulness in different situations. Knowledge can never be fixed at a single point but unfolds continuously as it is used, discussed, examined and so on. From this, it follows that knowledge is always in a state of flux, that is, process.

However, the theories and practice of knowledge to date have treated knowledge as a substance. Rather than substance, we should understand knowledge primarily as a process, produced and used in relation to the knowledge of other human beings who exist in relation to others. The value of knowledge relates to the effectiveness with which it enables systems to deal with their current activities and effectively envision and create their future.

All in all, knowledge is potentially useful information about something. Information is commonly represented by symbols. Symbols stand for or are about what is represented. Knowledge may be about what we call reality or about other knowledge. It is the implementation of stand systemizing for and about—the process of executing the epistemic cut—that (we need) to explore.

Chapter 6
Autopoiesis

Autopoiesis is the process whereby a system produces itself. This chapter provides a broad account of the developed form of autopoiesis theory. The account seeks to describe, in relatively straightforward terms, the various components and aspects of autopoiesis theory. Some of the key factors are the following.

In the 1950s and 1960s, open systems theory, together with sociological systems theory, was enormously influential in providing a coherent framework for the study of systems (i.e. organizations) and their environments. These approaches were in important respects motivated by ideals of ord/er, stability and predictability. They were so influential that the paradigm that they defined is still prevalent today. Although today's systems (i.e. organizations) and their environments are often characterized by transformation, emergence, considerable unpredictability and a strong emphasis on people, the systems thinking used to understand systems (i.e. organizations) is still not being conveyed in a coherent manner, for example to people leading business firms. The reason for this is the lack of a unifying framework for explaining a spectrum of systemic (i.e. organizational) phenomena from stable to highly dynamic systems and environments.

Therefore, this chapter elaborates some of the key ideas that shape the concept of this book. The overriding idea is that autopoiesis theory has the potential to provide a unifying framework for the study of systemic (i.e. organizational) phenomena in the twenty-first century. Although system (i.e. organization) studies have recently experienced no shortage of new paradigms and approaches—such as postmodernism, phenomenology, ethnomethodology, reflexivity and critical theory—the field seems to be expanding in ways that make it increasingly difficult to comprehend, especially for the uninitiated.

Autopoiesis is a concept developed through the pioneering work of Maturana and Varela (1980) in biology, primarily as a construct that enables a distinction to be made between living and non-living systems. The concept and its postulates have slowly been gaining ground and generating enthusiasm among many scientific communities. For Capra (1996), for example, Maturana and Varela's book *The*

K. U. Koskinen, R. Breite, *Uninterrupted Knowledge Creation*, SpringerBriefs in Business, https://doi.org/10.1007/978-3-030-57303-4_6

Tree of Knowledge (1992) contains no less than the outlines of two unified scientific conceptions of mind, matter and life. King (1993) suggests that autopoiesis is developing into a new theoretical paradigm in the social sciences, and von Krogh and Roos (1995a) suggest that autopoiesis offers the basis for a new general systems theory.

An extensive discussion of the connection of autopoiesis to cognition is provided by Thompson (2004). According to Thompson, the basic notion of autopoiesis as involving constructive interaction with the environment is extended to include cognition. Initially, Maturana (1988) defines cognition as the behaviour of an organism with relevance to the maintenance of itself. However, computer models that are self-maintaining but non-cognitive have been devised, so some additional restrictions are needed, and the suggestion is that the maintenance process, to be cognitive, involves readjustment of the internal workings of the system in some metabolic process. On this basis, it is claimed that autopoiesis is a necessary but not a sufficient condition for cognition. Thompson (2004, p. 127) takes the view that this distinction may or may not be fruitful, but what matters is that living systems involve autopoiesis and cognition as well. It can be noted that this definition of cognition is restricted and does not necessarily entail any awareness or consciousness by the living system.

In the emerging theory of living, a mind is not a thing but a process. It is cognition, the process of knowing, and it is identical to the process of life itself. This is the essence of the autopoiesis theory of cognition (Maturana and Varela 1980). The identification of mind (i.e. cognition) with the process of life is quite a new idea in science, but it is also one of the deepest and most archaic intuitions of humanity.

Furthermore, identifying cognition with the full process of life—including perceptions, emotions and behaviour—and understanding it as a process that involves neither a transfer of information nor mental representations of an outside world require a radical expansion of our scientific and philosophical frameworks. One of the reasons why this view of mind and cognition is so difficult to accept is that it runs counter to our everyday intuition and experience. As human beings, we frequently use the concept of information, and we constantly make mental representations of the people and objects in our environment.

Thus, an autopoietic system can be described as a random dynamical system, which is defined only within its organized autopoietic domain. Then, it is a modified definition of autopoiesis. An autopoietic system is a network of processes that produces the components that reproduce the network and that also regulates the boundary conditions necessary for its ongoing existence as a network. Hence, a system is cognitive if and only if sensory inputs serve to trigger actions in a specific way to satisfy a viability constraint. It follows from these definitions that the concepts of autopoiesis and cognition, although deeply related in their connection with the regulation of the boundary conditions of the system, are not immediately identical. A system can be autopoietic without being cognitive and cognitive without being autopoietic. A system that is both autopoietic and cognitive is a living system.

Thus, the theory of autopoiesis is presented as a biology of cognition (Di Paolo 2009). Cognitive science has been in need of theoretical grounding since its inception. This is not to say that cognitive science has lacked a theory. The representational view of the mind has dominated the field since its beginning and has developed to a very sophisticated degree. However, this view has been strongly criticized on different fronts (phenomenology, cognitive linguistics and situated artificial intelligence, to name a few). The general gist of these criticisms is that observing cognitive systems in the wild often produces a very different picture from that of cognitivism, one in which complex, causally spread processes encompassing the brain, the body and the environment self-organize in opportunistic ways to produce an appropriate performance under tight temporal constraints. These observations do not fit well with the representational picture, as they demand a deeper understanding of the autonomy and identity of a cognitive system (Di Paolo 2009). Hence, what some people have called the enactive approach has, at its most radical (Varela et al. 1991; Thompson 2007), turned to the theory of autopoiesis as its conceptual nucleus.

A system of the future needs an epistemology (i.e. a theory of knowledge) that is radically different from the epistemologies that have guided systemic thinking hitherto, and autopoiesis theory, with due adaptations, can furnish such an epistemology. In the following chapters, we begin by providing a brief overview of the key tenets of autopoiesis theory applied to systemic settings. Next, we discuss the system of the future, starting with the external pressures that are increasingly being exerted on social organizations of all types, inducing them to undertake new kinds of transformations. These chapters identify the important challenges facing systemic thinkers, now and in the foreseeable future, which exist not only as the result of the external pressures but also as a consequence of internal developments in system science and theory. The chapters conclude with an overview of the topics addressed by the contributors to this volume.

To summarize, the theory of autopoiesis thus belongs to the class of global theories, that is, theories that point to a collection of objects to which they themselves belong. Classical logic cannot really deal with this problem, and it will therefore be the task of a new systems-oriented epistemology to develop and combine two fundamental distinctions: between autopoiesis and observation and between external and internal self-observation. Classical epistemology searches for the conditions under which external observers arrive at the same results and does not deal with self-observation. Consequently, societies cannot be viewed, from this autopoietic perspective, as either observing or observable. Within a society, all observations are by definition self-observations.

6.1 Living Systems

To explain the nature of living systems, Maturana and Varela focus on a single, basic example—the individual, living cell. Briefly, cells consist of a cell membrane or boundary enclosing various structures, such as the nucleus and many molecules produced from within. These structures are in constant chemical interplay both with each other and, in the case of the membrane, with their external medium. It is a dynamic, integrated chemical network of incredible sophistication (e.g. Freifelder 1983; Alberts et al. 1989; Raven and Johnson 1991).

The biological evidence available today clearly shows that living systems belong to the class of autopoietic systems. To prove that the autopoietic system is the living system, it is then sufficient to show that an autopoietic system is a living system. This has been achieved by showing that, for a system to have the phenomenology of a living system, it suffices that its system is autopoietic (Maturana and Varela 1980).

Presently, however, it should be noticed that, in this characterization, reproduction does not enter as a requisite feature of the living system. In fact, for reproduction to take place, there must be a unity to be reproduced: the establishment of the unity is logically and operationally antecedent to its reproduction. In living systems, the system reproduced is the autopoietic system, and reproduction takes place through the process of autopoiesis; that is, the new unity arises in the realization of the autopoiesis of the old one. Reproduction in a living system is a process of division, which consists, in principle, of a process of fragmentation of an autopoietic unity with distributed autopoiesis such that the cleavage separates fragments that carry the same autopoietic network of production of the components that defined the original unity. Nevertheless, although self-reproduction is not a requisite feature of the giving system, it is just an occurrence. Living systems as we know them are a necessary condition for the generation of a historical network of successively generated, not necessarily identical, autopoietic unities, that is, for evolution.

In the original formulation of the autopoietic theory, Maturana and Varela talk about machines and not about systems. However, a machine can also be understood as a system, which is defined as any definable set of gears (Maturana and Varela 1980, p. 138). A more technical definition of a machine is a unity in the material space, differentiated by its organization, which implies a non-animistic point of view, the dynamism of which is noticeable (Maturana and Varela 1980, p. 136).

Maturana and Varela explain that a living being (which is also a unity) is characterized as continually self-producing and having an autopoietic system (Maturana and Varela 1980, p. 43). Therefore, the central idea of autopoiesis is that a living system is organized in such a way that all its components and processes jointly produce those self-same components and processes, thus establishing an autonomous, self-producing entity (Mingers 2010, p. 36). A clear example is the cell, because the cell produces each of its components through their interactions and transformations in a continuous way.

To sum up, the basic circular and self-referring nature of both physical living systems and nervous systems have an important evolutionary development.

Following from this, the identification of the primary unit of biology—the fundamental living system—is the individual rather than the species.

6.2 Autopoietic Systems

As noted by von Krogh and Roos (1995a), systems may have properties of autopoiesis and cognition. Varela et al. (1974), as discussed and quoted by von Krogh and Roos (1995a), list six criteria that are necessary and sufficient to recognize a system as being autopoietic:

1. The system is identifiably bounded (i.e. components of the system can be clearly distinguished from the rest of the world).
2. The system is comprised of a set of parts comprising an identifiable whole.
3. The system is mechanistic in that the component elements interact dynamically and/or act to control or transform one another. Properties of the system are generated by interactions between its components and are not simply the sum of the properties of those components taken individually.
4. The components forming the system's boundaries do so as a result of their interactions with other elements that identifiably belong to the system.
5. The elements forming the system's boundaries are produced by interactions of elements of the system, either by transformation of previously produced elements or by (catalytically) transforming and/or structurally coupling non-component elements that enter the system through its boundaries.
6. All other elements of the system are also produced by the interactions of its elements as in point 5 above and if those elements that are not so produced participate as necessary permanent constitutive components in producing other components.

If all these criteria are met, then the system is considered to be autopoietic.

That a system is autopoietic means that it only consists of self-producing elements. All the elements in the system are produced by the system itself through a network of such elements (Luhmann 1995a, p. 5).

As mentioned earlier, the theory of autopoiesis was developed by two Chilean cognitive biologists, Maturana and Varela, in the 1960s and early 1970s. They were trying to answer the question 'what is life'? In other words, 'what distinguishes the living from the non-living?' This question is similar but not identical to that of Monod (1974, p. 19), who is concerned with distinguishing between natural and artificial systems. Maturana and Varela conclude that a living system reproduces itself. They refer to this self-reproduction as autopoiesis. An autopoietic system is a system that recursively reproduces its elements through its own elements. Thus, autopoiesis means self-production, and an autopoietic system is a system that produces itself, that is, its components and its own boundary.

Moreover, autopoietic systems are more than self-sustaining in that they actually produce the components necessary for their own continuation. Such systems have properties such as autonomy, since they depend mainly on their own self-production and identity, maintaining their own individual autopoietic system despite changes in their structure.

An autopoietic system is organized as a network of processes of production of components that produces the components that, through their interactions and transformations, continuously regenerate and realize the network of processes that produces it and constitute it as a concrete unity in the space in which they exist by specifying the topological domain of its realization as such a network (Varela 1979). Thus, central to the concept of autopoietic systems is the idea that the different components of the system interact in such a way as to produce and reproduce the components of the system. This means that, through its components, the system reproduces itself.

All processes of autopoietic systems are produced by the system itself, and all processes of autopoietic systems are processes of self-production. In this sense, one can say that autopoietic systems are operatively closed. There are neither elements entering the system from outside nor vice versa. A system's operative closure, however, does not imply a closed-system model. It only implies closure on the level of the operations in that no operations can enter or leave the system. Nevertheless, autopoietic systems are also open systems. All autopoietic systems have contact with their environment. The contact with the environment, however, is regulated by the autopoietic system. The system determines when, what and through which channels energy or matter is exchanged with the environment.

This simultaneous openness and closure of the autopoietic system becomes particularly important when considering cognitive processes. For Maturana and Varela, the concept of living is directly linked to the concept of cognition. 'Living systems are cognitive systems, and living as a process is a process of cognition' (Maturana and Varela 1980, p. 13). In this sense, the operations of an autopoietic system are defined as its cognitions. Life and cognition are one and the same. Hence, everything that has been said about life applies to cognition. Cognition is a self-referential autopoietic process.

Therefore, the operative closure of the cognitive system means that the environment cannot produce operations in the system. Cognitions are only produced by other cognitions of the same system. The operative closure does not, however, imply a solipsistic existence of the system. In the opinion of Maturana and Varela (1980), operative closure is a precondition for interactional openness. On the level of its operations, the autopoietic system does not receive any inputs from the environment, only perturbations, which might trigger internal operations in the system. The particular processing of the perturbations from outside is entirely determined by the system itself (Mingers 1995).

Maturana and Varela generally distinguish between a system's organization and its structure. Organization refers to the interrelations between the components of the system, which define the system as a distinct system in a given space–time continuum. In this sense, the organization of the living system is autopoiesis. To speak of the same system, the organization of the system has to remain the same. That is, according to Maturana (1980b), the relations between the components that define a system as a composite unity of a particular kind constitute its organization. In this definition of organization, the components are viewed only in relation to their participation in the constitution of the whole that they integrate.

Structure refers to the actual components and the actual relations that these must satisfy in their participation in the constitution of a given composite unity and determines the space in which it exists as a composite unity that can be perturbed through the interactions of its components. However, the structure does not determine its properties as a unity (Maturana 1978). In contrast to the organization, the structure is not constitutive of the system. Structures can change, and one can still speak of the same system. Hence, the theory of autopoietic systems distinguishes strictly between the continuation of autopoiesis and the stabilization of particular structures.

Closely connected to the distinction between organization and structure is the shift from thinking in terms of structures to thinking in terms of processes. While conventional structuralist theories take their point of departure from the identification of structures and conceptualize processes as some sort of outcome of the structures, the theory of autopoiesis (e.g. Giddens 1990) starts off with processes and describes structures as their product. Although themselves dependent on structures, processes can be seen to be primary as they can produce new structures (e.g. Luhmann 2000).

6.2.1 Boundaries of Autopoietic Systems

Autopoietic systems must have a clearly identifiable boundary, like a cell membrane, independent of both the internal production network and the environment. This property, like some others, has been neglected or weakened in subsequent developments of the autopoiesis paradigm. However, it maintains a relevant role, because, without precise boundaries, a system's identity, invariance and closure become vague and uncertain, and the threshold between living and non-living systems is no longer an on/off state. Boundaries are identifiable and precise only over certain sensitivity thresholds.

In other words, to be considered as autopoietic, it is necessary for a system to have identifiable boundaries and to be capable of continually producing a boundary, but it is not necessary to have an explicit definition of the boundary or specific boundary elements. Mingers (1995) suggests that the components involved must create a boundary defining the entity, which means a whole interacting with its environment.

A boundary can also be defined as the fundamental distinction between the system and its environment, although the nature of the distinction can vary with time and location. According to von Krogh and Roos (1995a, p. 57), '... the boundary is created by individuals' knowledge pertaining to the organization–environment criterion. Each individual will form his or her own boundaries of the organization and recreate these dynamically as a part of their individual knowledge base.'

6.2.2 Unity, Organization and Structure of Autopoietic Systems

The basic cognitive operation, which the observers perform, is the operation of distinction. By means of this operation, observers specify a unity. When applying the operation of distinction to a unity so that the observers distinguish components in it, they specify it as a composite unity that exists in the space that its components define, because it is through the specified properties of its components that the observers distinguish it. If an autopoietic system is treated as a composite unity, it exists in the space defined by its components.

The relations between the components that define a composite unity (i.e. system) as a particular kind constitute its organization. In this definition of organization, the components are viewed only in relation to their participation in the constitution of the unity that they integrate. This is why nothing is said in it about the properties that the components of a particular unity may have other than those required by the realization of the organization of the unity.

The actual components, with their properties, and the actual relations between them that concretely realize a system as a particular member of the class of composite unities to which it belongs by its organization constitute its structure. Therefore, the organization of a system as the set of relations between its components that define it as a system of a particular class is a subset of the relations included in its structure. It follows that any given organization may be realized through many different structures and that different subsets of relations included in the structure of a given entity may be abstracted by an observer as organizations that define different classes of composite unities.

To understand better this relation between organization and structure, von Krogh and Roos (1995a) give a brief example—the organization of a bicycle requires two wheels connected by a frame. However, the structure of a bicycle may be modified be replacing wooden tyres with rubber tyres and a stainless steel frame with an aluminium frame. To understand autopoietic systems, however, we need to understand both the interrelations that define them and how the interrelations that constitute them are brought forth in the system.

The organization of a system, then, specifies the class of identity of a system and must remain invariant for the class of identity of the system to remain invariant. If the organization of a system changes, then its identity changes and it becomes a unity of another kind. Nevertheless, since a particular organization can be realized by systems with otherwise different structures, the identity of a system may stay invariant while its structure changes within limits determined by its organization. If these limits are overstepped—in other words, if the structure of the system changes so that its organization can no longer be realized—the system loses its identity and the entity becomes something else, a unity defined as another organization. Altogether, it is apparent that only a composite unity has both structure and organization.

6.2.3 *Structural Determinism and Coupling in Autopoietic Systems*

One of the main principles underlying the concept of an autopoietic system is that of structural determinism (Maturana 1991). As mentioned earlier, the organization that is realized with the help of the structure can interact and change. Then, so long as the structural changes maintain the organization, the system's identity remains. In other words, if the organization of a system changes, so does its identity.

All composite systems are structure determined. This means that the actual changes that the system undergoes depend on the structure itself at a particular instant (Mingers 2010). Any change in a composite system is a structural change and, as such, is determined by the properties of the components. Changes occur in response both to internal dynamics and to interactions with the environment, but, even in external interactions, the resulting change is determined internally. Then, a change is triggered by the environment. This is a very important conclusion, as it means that there can be no instructive interactions. In other words, it is never the case that an environmental action can determine its own effect on a structure-determined system.

Thus, the perturbations in the environment only trigger structural changes or compensations in systems. It is the structure that determines both what the compensations will be and what in the environment can or cannot act as a trigger. In total, the structure at any point in time determines:

- All possible structural changes within the system that maintain the current organization as well as those that do not
- All possible states of the environment that could trigger changes of state and whether such changes would maintain or destroy the current organization (i.e. identity).

Teubner (1991, p. 133) suggests that structure coupling at the level of systems can be defined in the following way: 'A system is structurally coupled to its environment when it uses events in the environment as perturbations in order to build up its own structure'.

Then, a central element of the theory of autopoiesis is the concept of structural coupling, which refers to the relation between systems and their environments. Environmental events can trigger internal processes in an autopoietic system, but the concrete processes triggered are determined by the structures of the system. That is, a system is said to be structurally coupled to its environment, or other systems in its environment, if its structures are in some way adjusted to the structures of the environment or the systems in the environment. The structures of the autopoietic system are not given but are themselves the result of the autopoietic reproduction of the system in its environment. It is important to note that, in the concept of structural coupling, adaptation is mutual. Not only is the system structurally adapted to its environment but also the other way around. This is particularly important if one considers the relation between two autopoietic systems, which constitute the environment for each other.

Moreover, structural coupling establishes a clear difference between the ways in which living and non-living systems interact with their environments. As Bateson (1972, 1979) points out, kicking a stone and kicking a dog are two very different stories. The stone will react to the kick according to a linear chain of cause and effect. Its behaviour can be calculated by applying the basic laws of mechanics. The dog will respond with structural changes according to its own nature and non-linear pattern of behaviour. The resulting behaviour is generally unpredictable.

As stated earlier, an autopoietic system is realized through a particular structure, and the changes that it can undergo are determined by that structure so long as autopoiesis is maintained. These changes may preserve the structure as it is or radically alter it. Where this is possible, the structure is said to be plastic (Mingers 1995). A plastic structure exists within an environment that perturbs it and can trigger changes. However, the environment does not determine the changes, but it can be said to select states from among those made possible at any instant by the system's structure. In an environment characterized by recurring states, continued autopoiesis will lead to selection in the system of a structure suitable for that environment. The system becomes structurally coupled to its environment and, indeed, to other systems within that environment.

Thus, the system is never idle. By observation, it distinguishes events in the environment, and it uses energy to discuss these events within the rules of its language. The system uses such events to discover new themes, issues, opportunities, threats, strengths and weaknesses. Gradually, new arguments are made that construct a description of the environment.

Every scientific inquiry begins with a distinction. The particular objects of interest are designated with respect to their internal operations, relations to other objects and so on. Hence, they are distinguished from objects of no or lesser interest, which are necessarily left unspecified in their respective operations, relations and so on (Spencer-Brown 1979, 1979, p. 3). The point of departure is the distinction between the system and the environment. A system emerges from boundary-producing operations that constitute an environment as everything the system is not (Luhmann 1984, p. 35). The environment is thus the negative correlate to the system. It is not life to living systems, it is not consciousness to consciousness systems and it is not communication to communication systems (Blaschke 2008).

The system and the environment are antidotes. The distinction between them is essential to both sides. There is no system without a corresponding environment and no environment without a boundary-producing system. The two emerge simultaneously from system operations. The environment is not the environment per se, just as there is not one objective reality. Each system constructs its own unique environment (Weick 1969, p. 63; Luhmann 1984, p. 249), its own unique reality (Berger and Luckmann 1966; von Foerster 2003).

6.2.4 Autonomy in Autopoietic Systems

Autonomy has been discussed in the literature since Aristotle, and several authors have used this term. Morin's (1982) essay on the intricate set of philosophical and scientific questions raises the fundamental principles of autonomy.

Autonomy means self-control, that is, maintaining identity. This means that a system is autonomous if it can specify its own laws for its own functioning (e.g. Morin 1982). This means that autopoietic systems are autonomous units. They subordinate all changes to the maintenance of their own autopoietic system. According to Varela et al. (1974, p. 188), 'Autonomy is the distinctive phenomenon resulting from an autopoietic organization. The realization of the autopoietic organization is the product of its operation.' Because an autopoietic system reproduces its own components and recreates its own system and identity, it acquires autonomy. The rules for its functioning are found in the system's organization and the way in which it reproduces itself. Autonomy is therefore a property of a living system in general: '... autonomy appears so obviously an essential feature of living systems that whenever something is observed that seems to have it, the naive approach is to deem it alive' (Varela 1979, p. 3).

Thus, the autonomy of an autopoietic system refers to its capability of being determined by its own internal rules instead of by inputs received from the environment. In such a context, the coupling between the system and its surroundings is given not by any input–output scheme but by a perturbation–dissipation effect, which is the direct consequence of the self-maintenance of the autopoietic system. In other words, to be cognitive means to be able to maintain a physical autonomous stability, despite the environment's constant perturbations (cf. von Krogh and Roos 1995a).

Autonomous systems are distinct from systems of which the coupling with the environment is specified or designed through input–output relations, like a computer (allopoietic system). In other words, an open organization of components and component-producing processes (linear or other noncyclical concatenations) leads allopoiesis. This means that the organization is not recursively generated through the interactions of its own products. In this sense, the system is not self-producing; it produces something other than itself. This allopoietic concatenation of processes is capable only of production, not self-production. Allopoietic organizations are still invariant and can be concatenated spontaneously (under favourable conditions).

Furthermore, allopoietic systems are organizationally open; they produce something different from themselves. Their boundaries are observer dependent, and their input and output surfaces connect them mechanically with their environment. Their purpose, as an interpretation of their input/output relation, lies solely in the domain of the observer.

A particular concatenation of production processes can be assembled by humans through a purposeful design. That is, man-made machines and contrivances with heteropoiesis, as well as their own productions, are heteropoietic. In other words, they are produced by another system. A machine, for example, is characterized by a system of components produced by other processes—a person or another machine—and of processes of production of which the products do not constitute the machine itself. So far, all heteropoietic systems are allopoietic (i.e. non-living).

It should be noted that the property of autonomy makes autopoietic systems distinct from self-organizing systems (Jantsch 1980; Andrew 1989). They differ with respect to the criteria of the autonomy that they imply. Systems first have to be self-organized before they can become autopoietic. In other words, autopoiesis is not synonymous with self-organization, as suggested by some authors (e.g. Zimmerman and Hurst 1993).

6.2.5 Observer and Observation in Autopoietic Systems

'If there is anything like a central intellectual fascination in this century it is probably the discovery of the observer' (Baecker 1996, p. 17).

Spencer-Brown suggests treating observation as the most basic concept for any analysis. As a concept, it is supposed to be even more basic than that of a thing, event, thought, action or communication. This means that the term observation is not used in its usual sense as referring merely to optical perception. Instead, observation is used as an abstract concept referring to any operation from communication to thought and even to the operation of a machine. Even the observer is treated as an observation (Spencer-Brown 1979).

However, the role of an observer is often ignored in systems writing (Weinberg 2001). The most popular way of ignoring the observer is to move straight into a mathematical representation of a system—a so-called mathematical system—without saying anything about how that particular representation was chosen. For example, Hall and Fagen (1968, p. 81) provide this definition: 'A system is a set of objects together with relationships between the objects and between their attributes'. These authors rightly emphasize relationships as an essential part of the system concept but fail to give the slightest hint that the system itself is relative to the viewpoint of an observer.

All descriptions and explanations are made by observers who distinguish entities or phenomena from the general background. That is, everything that is said is said by an observer. Such descriptions will always depend in part on the choices and purposes of the observer and may or may not correspond to the actual domain of the observed entity (Maturana and Varela 1980). That which is distinguished by an observer Maturana calls a unity, that is, a whole distinguished from a background. In other words, in making the distinction, the properties that specify the unity as a whole are established by the observer. The result will be a description of a composite unity in terms of components and the system that combines the components together into a whole.

One must not confuse that which pertains to the observer with that which pertains to the observed. Observers can perceive both an entity and its environment and see how the two relate to each other. Components within an entity, however, cannot do this but act purely in response to other components (Mingers 1995).

Thus, the observer can choose to focus his or her attention either on the internal structure of the system or on its environment. In the former case, the observer sees the environment as background and the properties of the system emerge from the interaction between its components. In the latter case, the observer treats the system as a simple entity with certain interaction with the environment. This means that a system is a way of looking at the world.

Every observation is construed from two components: a distinction and an indication. An observer chooses a distinction with which he or she demarcates a space into two spaces (states or contents). Of these two states, he or she has to choose which he or she indicates. That is to say, the observer has to focus on one state while neglecting the other. It is not possible to focus on both simultaneously. In this sense, the relation between the two states is asymmetrical (e.g. Cooper 1986; Chia 1994).

Furthermore, the concept of observation does not focus on the object of observation but on the observation itself as a selection of what to observe (Latour 1986). In this sense, the underlying question concerns not what an observer observes but how an observer observes: how is it that an observer is observing what he or she is observing and not observing something else?

The real world gives the product space of what exists. However, that product space represents the uncertainty of the observer. The product space changes if the observer changes, and two observers may legitimately use different product spaces within which to record the same subset of actual events in some actual thing. The constraint is thus a relation between an observer and a thing. The properties of any particular constraint will depend on both the real thing and on the observer. From this, it follows that a substantial part of the theory of product space will be concerned with properties that are not intrinsic to the thing but are relational between the observer and the thing (Ashby 1968).

Because the autopoietic process is not directly accessible to anything or anybody except the system (i.e. it is only open to observation), any characterization of an autopoietic system can only be given from the standpoint of an observer (von Foerster 1972). An observer, or observer community, is '... one or more persons who embody the cognitive point of view that created the system in question, and from whose perspective it is subsequently described' (Varela 1979, p. 85).

Thus, the fundamental cognitive operation that an observer performs is the operation of distinction. An operation of distinction, however, is also a prescription for a procedure, which, if carried out, severs a unity from a background, regardless of the procedure of distinction and regardless of whether the procedure is carried out by an observer or by another entity.

Furthermore, the prescription of an operation of distinction implies a universal phenomenon of distinctions, which, through the specification of new procedures of distinction or through their recursive application in the re-ordering of the distinguished entities, can, in principle, endlessly give rise to phenomenal domains.

6.2.6 Summary

Autopoietic means that a system consists only of self-producing elements. All elements in the autopoietic system are produced by the system itself through a network of such elements. Moreover, an autopoietic system is more than self-sustaining in that it actually produces the components necessary for its own continuation. Such a system has properties like autonomy, since it depends mainly on its own self-production and identity and maintains its own individual autopoietic system despite changes in its structure.

Simultaneous openness and closure of the autopoietic system are particularly important when considering cognitive processes. Living is directly linked to the concept of cognition.

Maturana and Varela generally distinguish between a system's organization and its structure. Organization refers to the interrelations between the components of the system, which define the system as a distinct system in a given space–time continuum. Structure refers to the actual components and the actual relations that these must satisfy in their participation in the constitution of a given composite unity and determines the space in which it exists as a composite unity that can be perturbed through the interactions of its components.

An autopoietic system requires identifiable boundaries and is capable of continually producing boundaries, but it does not require an explicit definition of the boundaries or need specific boundary elements.

The organization of an autopoietic system specifies its identity and must remain invariant for the class of identity of the system to remain invariant. If the organization of a system changes, then its identity changes and it becomes a unity of another kind.

A central element of the theory of autopoiesis is the concept of structural coupling, which refers to the relation between systems and their environments. Environmental events can trigger internal processes in an autopoietic system, but the concrete processes triggered are determined by the structures of the system. Then, an autopoietic system is structurally coupled to its environment when it uses events in the environment as perturbations to build up its own structure.

The autonomy of an autopoietic system refers to its capability of being determined by its own internal rules instead of by inputs received from the environment. In such a context, the coupling between the system and its surroundings is not given by any input–output scheme but by a perturbation–dissipation effect, which is the direct consequence of the self-maintenance of the autopoietic system.

Knowledge depends very much on the point of observation of an individual. In autopoiesis theory, knowledge and observation are closely related, since observing systems are autopoietic systems (cf. Piaget 1936). To be more precise, in autopoiesis theory, distinctions and norms are two central categories (Varela 1979; Luhmann 1986, 1988). Knowledge is what makes individuals able to make distinctions in their observations and, based on their norms, determine what they see. The distinctions reveal the knowledge of the observer (Koskinen 2013).

Chapter 7
Social Autopoietic Systems

Niklas Luhmann developed an elaborate theory of social and cognitive systems, which combines Maturana and Varela's notion of cognition with Husserlian phenomenology (Arnoldi 2006, p. 117). That is, Luhmann developed the idea of autopoiesis of social systems and recognized that this use of autopoiesis could be problematic (Luhmann 1986, p. 172). As humans are central elements of social systems, it follows that, to be considered as an autopoietic system, a social system must be self-reproducing in terms of humans (Bednarz 1988, p. 61). How could this be possible? Luhmann proposed a very intelligent solution to this paradox. He redefined social systems as being realized in a domain of communication. In other words, the constituent elements of a social system are communications. Therefore, the social system is understood as a network of communication that emerges over time (Nassehi 2005, p. 181). Subsequently, the conditions for autopoiesis have to be evaluated in terms of the self-production of communication (Teubner 1991, p. 235). In fact, Luhmann understands communication not as a diffusion of meaning or information from one person to another but as an autopoietic system that appears out of the doubly dependent meeting of subjects (Arnoldi 2006, p. 116). As a corollary, we can consider that Luhmann's theory is an autopoietic theory of communication (Mingers 2010, p. 158).

This means that Niklas Luhmann drew on the new theoretical ideas from systems theory of autopoiesis to produce '... the most developed and most radical attempt within contemporary sociology to recast completely the theory of society' (Harrison 1995, p. 65).

Luhman suggests that we speak of autopoiesis whenever the elements of a system are reproduced by the elements of the system (Luhmann 1992, p. 128). This criterion, as he points out, is also met by non-biological systems. Apart from living systems, Luhmann identifies two additional types of autopoietic systems: social systems and psychic systems. While living systems reproduce themselves on the basis of life, social systems reproduce themselves on the basis of communication and psychic systems on the basis of consciousness. Furthermore, living systems can be

K. U. Koskinen, R. Breite, *Uninterrupted Knowledge Creation*, SpringerBriefs in Business, https://doi.org/10.1007/978-3-030-57303-4_7

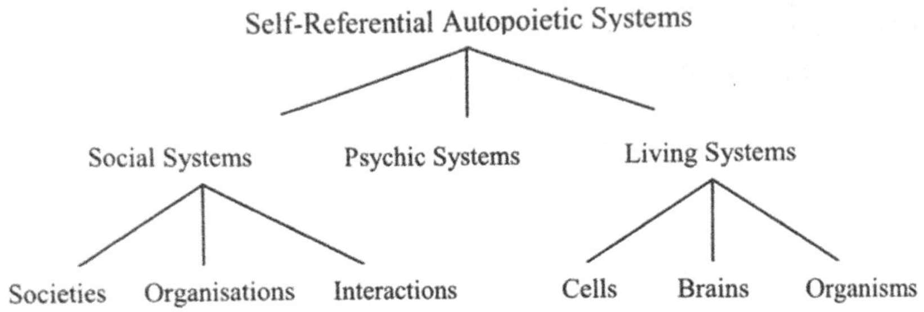

Fig. 7.1 Types of autopoietic systems (Source: Luhmann 1986, p. 173)

differentiated into three subsystems: cells, brains and organisms. Equally, social systems can be divided into three subsystems: society, interaction and organization (Fig. 7.1).

On the basis of this typology of systems, one can derive a hierarchy of three levels of analysis (Luhmann 1986). On the first level, we find statements that concern autopoietic systems in general without reference to any particular mode of reproduction. On this level, we can find the general concept of autopoiesis (Chap. 6). Statements on this level are equally valid for living and for social and psychological systems.

On the second level, we find different applications of the general theory of autopoiesis. There are three such areas: research concerned with the particular characteristics of social systems, psychic systems and living systems. Most of Maturana's and Varela's research can be placed on this level (in the area of living systems). It produces general statements concerning all types of living systems (cells, brains and organisms) but is not applicable to social or psychic systems. Psychological research is concerned with the particularities of systems that are reproduced on the basis of consciousness. Sociological research on this level is concerned with the particularities of systems reproducing themselves on the basis of communication. Statements produced in this area concern all three types of social systems.

On the third level, there is research in the social field dealing with the particularities of societies, systems (e.g. business companies and interactions). That is, for each type of system, the particular mode of production has to be defined and the consequences of the particular mode of production analysed. Thus, for social research in particular, there are four different areas of research, namely research on the general level of social systems and research on the particular types of social systems: on societies, on systems (e.g. business firms) and on interactions (cf. Seidl and Becker 2005).

Thus, in social research in particular, one can identify four different areas: research on the general level of social systems (e.g. Luhmann 1995a) and research on the particular types of social systems—on societal systems (e.g. Luhmann 1997), on systems like business firms (e.g. Luhmann 2000) and on interactional systems (e.g. Luhmann 1993, pp. 81–100) (Fig. 7.2).

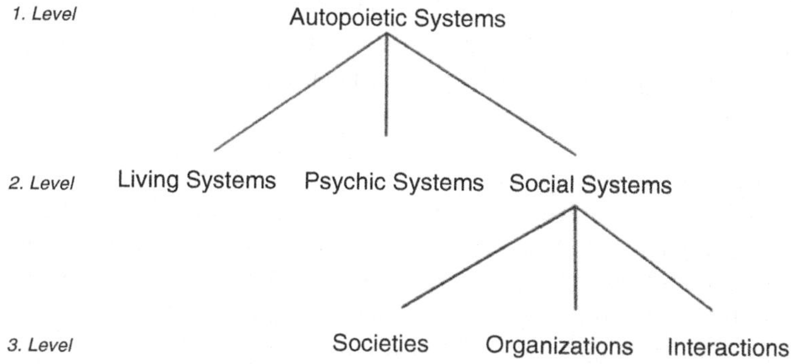

Fig. 7.2 Social autopoietic systems (Source: Luhmann 1986; Seidl and Becker 2005)

Against the backdrop of the categorization of analytical levels, the transformation of the original autopoiesis concept into the concept of autopoiesis of particular types of social systems becomes clear. Instead of just transferring the concept from the field of biology to the field of sociology, it is first abstracted to a general concept on a transdisciplinary level before being respecified into social autopoiesis and the autopoiesis of particular types of social systems.

Thus, Luhmann's general concept of autopoiesis radicalizes the temporal aspect of autopoiesis. While Maturana and Varela conceptualize the elements of biological systems as relatively stable chemical molecules, which have to be replaced from time to time, Luhmann conceptualizes those elements as momentary events without duration that vanish as soon as they come into being; they are momentary and immediately pass away (Luhmann 1995a, p. 287).

This is to say, through this shift from a reproduction of relatively stable elements to a reproduction of momentary events, Luhmann radicalizes the concept of autopoiesis. Because the elements of the system have no duration, the system is urged to engage in the constant production of new elements. If the autopoiesis stops, the system disappears.

In addition to temporalization, Luhmann de-ontologizes the concept of elements. Elements are defined as such merely through their integration into the system. Outside or independently of the system, they have no status as elements; that is to say, they are not ontologically pre-given (Luhmann 1995a, p. 22). Elements can be composed of different components, which can be analysed independently of the system, but, as elementary units, they are only defined through the functions that they serve for the system as a whole (Luhmann 1997, p. 66).

As a consequence of ontologizing the concept of elements, the concept of production (as in self-reproduction) obtains a functional meaning. Production refers to the use of an element in the network of elements. The important point in this conceptualization is that the element and the use of the element are not two different coins but two sides of the same coin. It is not that we first have the element and then the system makes use of it; only by being used (i.e. by becoming integrated into the network of other elements) does an element become an element. Thus, one can say that the element is produced as a result of being used (Luhmann 1997, pp. 65–66). One can, of course, analyse the substratum on which an element rests and find a whole range of causal factors that are involved in bringing it about, but the particular unity, as which the element functions in the system, can only be produced by the system.

As stated earlier, through this shift from the production of relatively stable components to the production of momentary events, Luhmann radicalizes the concept of autopoiesis. Because the components (i.e. events) of the system have no duration, the system is urged to produce new components (i.e. new events) constantly. If the autopoiesis stops, the system disappears immediately.

According to the original definition (i.e. Maturana and Varela 1980), autopoietic systems interact with themselves rather than with their environments. However, based on Maturana's (2002) later definition, it is clear that real autopoietic systems are physical systems that are entropically dissipative and must remain open to fluxes of matter and energy to fuel their self-production. In Luhmann's version of autopoiesis, systems interact cognitively with their own interpretations of the external world rather than with the external world per se. Interpretations are formed over time; hence, systems are formed by their own historicity as past choices and the effects of those choices interact to constitute the system's identities.

It is crucial to understand that autopoietic systems are not more or less open to their environments, as assumed in traditional systems theory; rather, they are both closed and open. They are closed for their own operations, allowing interaction with themselves, and they are open for observation of the outside world. In processes of communication, for example, an individual (receiver) is not seen to receive parcels of data. Instead, the communication takes place as an individual interacts with his or her own cognitive framework while remaining aware of what has been uttered by the other individual (sender) (cf. Hernes 2008).

Furthermore, Luhmann takes quite a different view of the ontological status of the system. For functionalists, the system is taken for granted as a stable, well-ordered entity in an ongoing open interaction with its environment, with boundaries that are clearly defined even for the external observer. For Luhmann, the picture is quite different. The system is in a continual process of recreating itself anew as communications generate further communications and thereby an internal structure. It is the system that defines its boundaries through its communications and thereby closes itself off, generating quite an impermeable barrier to the environment. It is interactively open, but the effects of events in the environment of the system (if any) are determined by the system itself (Seidl and Becker 2005).

The idea that social systems are continually creating their own components and their own boundaries makes them inevitably self-referential. They are able to draw a distinction between themselves and their environment and to represent and enact the distinction in their operations. It also means that social systems exhibit a degree of closure in at least two senses: first in being able to accept or reject that which does or does not belong to the system and second in that operations of a particular type only generate operations of the same type—communications generate communications, meaning generates meaning and thought generates thought. These general characteristics of self-construction, self-reference and closure lead Luhmann to adopt the theory of self-producing systems, that is, autopoiesis.

Thus to summarize, organization theorists, whether they come from functionalist or interpretive schools of sociology, acknowledge the existence of systems, albeit in different ways. Luhmann's radicalism lies in two assumptions about systems. First, systems are real; hence, systems theory is not a mere analytical means of analysing the social world. Systems belong to the real world, because, without them, the social world disintegrates, as it is no longer possible to distinguish what is from what is not: what is the system and what is the environment. This is to say, systems exist. Second, systems are autopoietic, that is, self-referencing. To exist over time, they need to be able to reproduce themselves, which means that they must reproduce meaning, just as actions must reproduce actions. Their reproduction takes place through their connecting of operations over time, which enables them to interact with their own processes of creating meaning. 'Everything that is used as a unit by the system is produced by the system itself. This applies to elements, processes, boundaries, and other structures, and last but not least, to unity of the system itself. Autopoietic systems, then, are sovereign with respect to the constitution of identities and differences' (Luhmann 1990a, p. 3).

7.1 Events

Events constitute a major aspect of Luhmann's autopoiesis, notably through their role in creating the system's temporality. Events take place in time, and they mark the difference between before and after. Events exist in time but have by themselves no extension in time, that is, they are essentially evanescent phenomena. They are moments in time that exist as markers that allow us to explain the continuity as well as the discontinuity of social systems. Events mark selections of some alternatives over others. If an unfortunate decision is made, correcting it will not eliminate that decision, but its correction will enter the process as a new event.

An important feature of events—and hence processes—as opposed to that of structures is that they are irreversible. Once an event has taken place, it becomes a fact as having taken place. Events occur only once and only in the briefest period necessary for their appearance, which makes them ideal as elementary units for the analysis of processes (Luhmann 1995a, p. 67). Events, compared with structures, do not keep options open but mark irreversibly the choice of some options over others. Because systems depend on being both reversible and irreversible (keeping options open while also making selections), both structure and process become important to systems. A structure without a process becomes an empty cell; a process without a structure becomes actions without direction. Therefore, Luhmann does not treat processes as isolated from structures but considers structure and process as complementary terms (Luhmann 2000, p. 340). Structures and processes become related to one another through events, during which selections are made.

Furthermore, according to Luhmann, social systems consist of recursively connected elements. These elements are seen as communicative events, not as actions or individuals. Communication, however, is only really communication if the communicative event is understood and '. . . used as the basis for connecting with further behaviours' (Luhmann 1995a, p. 141). Thus, communication comes into being as communication only when communicative events consecutively connect to each other.

Thus, as stated above, events take place in time and mark the difference between before and after. Whereas Weick (1979) also emphasizes the importance of events in organizing processes, Luhmann develops the notion of events further.

First, events are irreversible. Once they have taken place, they cannot be undone. Second, events exist in time but have by themselves no extension in time. Rather, they exist as markers that allow us to explain the continuity as well as the discontinuity of social systems. In Luhmann's organization (system) theory (Luhmann 2000), for example, decisions act as events marking the difference between before and after (Åkeström Andersen 2003). Decisions form points along a time continuum: points that themselves have no extension. Third, events mark transitions. However, they are not mere neutral transitions in which one course of events stops and another takes over. Events carry with them potentiality, implying that something is possible after the event that was not possible before it.

Furthermore, it is by focusing on events that we open up the study of not just the continuity of a system but also its discontinuity. Many studies in organization (system) theory either may be causal in the sense that explanations are found for why changes take place in a certain way or may reject any causality at all; Luhmann's autopoiesis carries with it an inbuilt ontological insecurity in the sense that we study that which takes place, but we may equally well study that which does not take place. An example could be that of studying trust in organizations. Trust, in Luhmann's work, is not a continued state, such as what seems to be implied by Giddens (1990). Instead, we should always reckon with the possibility of mistrust, because there is always the possibility that the world that we take for granted is turned upside down. Continuity may turn into discontinuity, just as trust may turn into mistrust. This is why processes need to be supplemented by events. Processes largely express continuity while events allow for discontinuity. Luhmann (2000, p. 46) draws on a formulation by Ingold (1986, p. 24) to illustrate the relationship between event and process: 'Process is to event as continuity is to discontinuity'.

To sum up, articulation serves to bring meanings to life, which in turn provides acts with meaning. Meanings are constituted by accumulations of events over time; hence, they may be seen as event-objects. Events provide meanings with historicity and direction.

7.2 Language

Von Krogh and Roos (1995a) make one of the most significant contributions to integrating autopoiesis into management theory and research. In so doing, they advance an anti-cognitivist position in the systemic knowledge debate. They reject the notion that knowledge is a given and that the task of systems is to represent it as accurately as possible. Instead, they argue that knowledge is embodied in human beings and that representations of the world in the human mind come forth as a result of actions or observations by human beings. 'Knowledge is what brings forth a world' (von Krogh and Roos 1995a).

Language is an enigmatic phenomenon. This is clear from the controversy surrounding it. At one end of the spectrum, we have those who see language as a mere pointing tool that humans use to point at the world or to act as containers to make external that which is internal, such as thoughts, ideas, self and so forth. This is what Taylor (1985) refers to as the designative view of language—also sometimes called the denotative role of language. In this view, language is a rather unproblematic system of pointers that merely stand for things, thoughts, feelings and so forth. The meaning of the word is what it designates. In this view, the problem of language is to sort out or unravel the system of signifiers, and that which they signify, by means of more precise unambiguous descriptions/definitions and the rules of logic. This is by no means a trivial task for them; however, they see it as always possible in principle.

The relationship between language and the world has two ontological dimensions. The first dimension is language, as the historical already-there horizon of the world. Language started with the first word and continues to unfold as an ever-expanding horizon of meaning—the already-spoken word, not in itself but as the collective world of community. It is in the world, in this community, that we speak when we speak. In speaking/listening in the world, we open up possibilities for unexpected meanings to emerge. It is these possibilities for expressing meaning that make us stand out as significant others. It is these two dimensions of language, as the already-there horizon and as original expression, that we need to explore as the essential dynamic of social autopoiesis.

In the autopoiesis theory, self-awareness is viewed as being tied closely to language, and the understanding of language is approached through a careful analysis of communication. Communication, according to Maturana, is not a transmission of information but rather a coordination of behaviour among living organisms through mutual structural coupling. Such mutual coordination of behaviour is the key characteristic of communication for all living organisms, with or without nervous systems, and it becomes more and more subtle and elaborate, with nervous systems of increasing complexity. Thus, linguistic communication requires a nervous system of considerable complexity, because it involves a considerable amount of complex learning (Maturana 1970).

Our linguistic distinctions are not isolated but exist '... in the network of structural couplings that we continually weave through languaging' (Maturana and

Varela 1980, p. 234). Meaning arises as a pattern of relationships among these linguistic distinctions, and thus we exist in a semantic domain created by our languaging. Finally, self-awareness arises when we use the notion of an object and the associated abstract concepts to describe ourselves. Thus, the linguistic domain of human beings expands further to include reflection and consciousness.

The uniqueness of being human lies in our ability to weave continually the linguistic network in which we are embedded. To be human is to exist in language. In language, we coordinate our behaviour, and together in language we bring forth our world. This human world centrally includes our inner world of abstract thought, concepts, symbols, mental representations and self-awareness. To be human is to be endowed with reflective consciousness: 'As we know how we know, we bring forth ourselves' (Maturana and Varela 1980, p. 244).

As pointed out by Wittgenstein (2009), the usage of words plays an important role in the way in which people communicate with each other. Particular usages of words tend to be specific to national cultures, to regional sub-groups within a nation as well as to systems like business firms.

It is in the systemic tradition (like in firms) of languaging to make particular use of words. Wittgenstein suggests that words are embedded in so-called language games. They derive their meaning from the content of their use rather than the objects, events or actions that they denote. The use of words follows certain history-dependent rules that are specific to an institutional setting. Such rules are created and recreated in languaging and form the basis for the social system's knowledge of the world. The very uses of the double distinction making of languaging carry, in themselves, their own distinguishing capacity. Thus, the very use of a word in a particular way is distinct from a different way of using the word (von Krogh and Roos 1995a).

The language game has many functions. First, it helps to conserve the system's (like business firms') knowledge system. Each system carries its own unique set of rules, distinct from other systems and other rules that provide a distinct system of meaning. The system provides some coherence for the usage of words for system members (Eisenberg 1984; Astley and Zammuto 1992), and this coherence allows system members to distinguish the right use of words from the wrong ones. As such, the language games of the system provide a template from which to coordinate meanings and subsequently interpret and coordinate action. For example, to give orders or to obey orders, system members must use words like action, policy, time, control, perform and so on to allow for the initiation of action and observation as well as the description of action once it occurs.

A normal part of systemic languaging, unavoidable in a world in which the knowledge of the individual is embodied, is to contest, challenge and question rules. Rules are subject to innovation as system members make new uses of words. Frequently people need to innovate in the way in which they state their policies. Difficult issues are settled in an ambiguous form that allows multiple interpretations by various groups of system members (Meyer 1984; Astley and Zammuto 1992). Issues become defined not by a reference to particular objects, events or actions but by the way in which they are used in the system. What must be understood with coherence, thus, is that the system always provides a context for playing new language games. There are always starting points from which new words and rules for the uses of words can be developed.

The rules help to decrease the time needed for conversation to initiate competent action or further conversations. In an example given by Wittgenstein (2009; von Krogh and Roos 1995a), a builder is building with stones: there are blocks, pillars, slabs and beams. An assistant has to pass him the stones in the order in which the builder needs them. For this purpose, they make use of a language consisting of the words block, pillar, slab and beam. The builder calls them out; the assistant brings the stone, which he has learnt to bring in response to such-and-such a call. The assistant does not have to hear, by contrast, the assistant: could you give me the stone, frequently referred to as a block that is situated at the storehouse of stones, please. The function of socialized system knowledge is to allow for rules and languaging that lead to effective action.

Linguistic behaviour is connotative. The observed communication of meaning and practical efficacy of language do not reside in the words and terms themselves but reflect similarities in the organisms' structures developed through their history of interactions (Mingers 1995).

Organisms that interact repeatedly with one another become structurally coupled. They develop behaviours that reciprocally trigger complementary behaviours, and their actions become coordinated to contribute to the continued autopoiesis of each. The particular behaviours are divorced from what they connote; they are symbolic and thus essentially arbitrary and context dependent. They only work to the extent that they reflect agreement in structure, and this is what Matura means by a domain of consensual action. They rely on consensuality rather than explicit consensus among those involved (Harnden 1990).

Thus, language does not transmit information, and its functional role is the creation of a cooperative domain of interactions between speakers through the development of a common frame of reference, although each speaker acts exclusively within his or her cognitive domain in which all ultimate truth is contingent on personal experience. Since a frame of reference is defined by the classes of choices that it specifies, linguistic behaviour cannot but be rational, that is, determined by relations of necessity within the frame of reference within which it develops. Consequently, no one can ever be rationally convinced of a truth that he or she did not have already implicitly in his or her ultimate body of beliefs (Maturana 1970). Thus, knowledge travels through language. Language is the verbal blueprint of our experience. Without a word or a language to describe our experience, we cannot communicate what we know. Every mode of knowledge travels through a different language. Language initiates us into a particular world of experience. Expanding our systemic knowledge means that we must expand the language that we use to describe our systemic experience.

To summarize, language and languaging are useful interpretative lenses through which to understand different knowledge creation processes, because they act as devices for systems: to make sense of past events and actions (Reissman 1993); to present themselves and others (Horrocks and Callahan 2006); to share meaning in a collectivity (Ylijoki 2005); and to provide legitimacy and accountability for their actions (Czarniawska 1997; Currie and Brown 2003).

7.3 Metaphors

A metaphor is an assertion that A is M or that A is like B (Easton and Araujo 1993). For example, one might say that life is just a bowl of cherries or an atom is like the solar system (Koskinen 2013). It is important to recognize that a metaphor, as a figure of speech, is not simply an object: it expresses a relationship. To capture this relationship, individuals use the terminology of the base and target domains. A metaphor implies that the target domain is like the base domain. The similarity between the domains is a crucial aspect of the process of using metaphors. Then, metaphors in language are more than simply literary devices. They are central to knowledge creation processes and ways of knowing as well as being ubiquitous in everyday language.

A metaphor consists of two main parts: the tenor and the vehicle. The tenor is the subject to which the metaphor is applied. The vehicle is the metaphorical term through which the tenor is applied. These two parts come together to reach a point of similarity known as ground. For a metaphor to work best, there has to be just enough distance between the tenor and the vehicle for an individual to be able to make some connection between these two, but they should not be so close as to appear over obvious or so far away as to appear obtuse (Parkin 2003).

Indeed, metaphors are terms that cast light on a phenomenon by virtue of association with something familiar to us. Morgan (1996) states about metaphors that, although they may be regarded as devices of embellishment, their significance is in fact much greater. The use of metaphors, in Morgan's view, implies ways of thinking and seeing that pervade how we understand our world generally. Morgan (1996, p. 13) argues further that '. . . many of our taken-for-granted ideas about systems are metaphorical, even though we may not recognize them as such'.

Thus, linguists propose that words are introduced into a language whenever it becomes desirable to make functionally important distinctions in a given context of human endeavour (Bickerton 1993). For example, externalizing tacit knowledge into explicit knowledge means finding a way (e.g. a word) to express the inexpressible. One of the means of doing so is the usage of figurative language and symbolism—metaphors (e.g. Lakoff and Johnson 1980; Tsoukas 1991). Using metaphors is a distinctive method of perception. It is a way for individuals grounded in different contexts and with different experiences to understand something intuitively through the use of imagination and symbols without the need for analysis or generalization (von Krogh and Roos 1995b). Metaphors are a special kind of meanings in a person's worldview. Through metaphors, people put together what they know in new ways and begin to express what they know but cannot yet say. As such, metaphors are highly effective in fostering direct commitment to the creative process in the early stages of knowledge creation.

When people recognize the idea—albeit unconsciously—that is given by a metaphor and make the connection, their own understandings are enhanced and they discover new ways of dealing with the problem. This means that recognizing the idea serves as an interruption, which, in turn, means that it gives individuals a 'what?' sort of experience and jolts them out of their habitual thinking (Denning 2001). Thus, a metaphor can also bypass individuals' natural resistance to change. If individuals perceive that they are being told what to do or given advice, then there is always a danger that protective barriers will arise. Being offered possible solutions through a metaphor is more acceptable and non-threatening, and there is less perceived pressure to accept the advice being offered.

For example, a successful firm operating in a rather stable business environment could describe itself as a river. This metaphor defines the firm as a system that moves at a fairly slow pace but is consistent and achieves things. However, when the business environment changes to turbulent, and therefore there is a need to make the firm's functioning philosophy more dynamic, the management could describe the firm with the metaphor of a chameleon. This metaphor describes the firm as a system that has the ability to change its appearance rapidly in response to threats and opportunities in its business environment (Koskinen 2010a, b).

To sum up, metaphors link ideas so that new knowledge can emerge. They are relevant to any situation that requires both new thinking and reflection on past experience. In the complex world, the impossibility of comprehensive description is resolved by making sense through pattern recognition. Metaphors provide the language to describe those patterns and communicate their meaning. We need to remember that language creates reality, so the metaphors that we adopt generate the perception of reality for ourselves and for the people with whom we are communicating. In other words, a metaphor can merge two or more different and distant areas of experience into a single, inclusive image or symbol, which Black (1962, p. 38) aptly describes as 'two ideas in one phrase'. By establishing a connection between different things that seem to be only distantly related, metaphors set up a discrepancy or conflict.

7.4 Decision and Decision Maker

Decision theory is multidisciplinary and treats all aspects of choice. It is the foundation of the behavioural and social sciences. Decision theory examines and refines decision theory's philosophical claims. Its primary subject is rational choice. Thus, it deals with normative matters and is allied with branches of philosophy such as epistemology and ethics. This means that decision theory assists epistemology in its study of rational belief and assists ethics in its study of good acts, goals and character traits. The behavioural and social sciences use decision theory to construct models of human behaviour. Often a theory of rationality yields a good first approximation of human behaviour. A general theory of rationality covers individuals and groups of people and shows how the rationality of individuals leads to the collective rationality of groups. It offers a fruitful foundation for the behavioural and social sciences. Then, decision theory's main normative question concerns the nature of rationality. Which principles of rationality govern choice? Principles of instrumental rationality counsel the adoption of means appropriate to one's ends. A large body of literature explicates this advice. It elaborates the view that instrumental rationality requires the maximization of subjective utility: that is, acting to maximize the achievement of one's goals. An instrumentally rational act is rational conditional on the rationality of the ends that the act serves and the agent's processing of evidence about appropriate means. Other principles of rationality govern the adoption of ends.

Luhmann argues that the central element around which any viable concept of a system has to be built is a decision. Nevertheless, the concept of a decision generally seems to be even less clear than that of a system (e.g. Mintzberg and Waters 1990). Mostly a decision is defined as a choice. This means, however, that a decision is defined through a synonym that is equally unclear. Sometimes the definition is specified somewhat more, describing it as a choice among alternatives. The concept of an alternative, however, is itself only defined in relation to a choice. Alternatives are those possibilities among which one can choose—the choice defines the alternatives (Luhmann 2000, p. 125). Thus, we have a tautological definition, which in this form does not really help very much. To arrive at a more fruitful concept, Luhmann argues that one has to unfold the tautology and analyse its particular form. As he demonstrates, this analysis can be performed both from a factual and from a temporal perspective.

The decision maker can be approached and understood in three dimensions (Pihlanto 2009). Consciousness makes the inner world of the decision maker problematic. The functioning of the consciousness denotes a complicated and constant act of understanding, which is realized in terms of meaning. Understanding phenomena means knowing, feeling, believing, intuiting and so on. The act of understanding is highly individual and subjective, and such are also the resulting contents of the worldview of an individual. The subjective worldview of the decision maker forms a personal historical basis for understanding. All this accentuates the subjective nature of human cognition and action.

The situationality of a decision maker means that a multiplicity of situational factors and their combinations are reflected in the consciousness—not only single and concrete or factual—but also ideal or intangible ones. Situation and situationality are individual and subjective, similar to the process of forming meanings in the consciousness. All this means that every decision maker is unique in terms of his or her way of understanding and acting in an action context (Pihlanto 2009).

Finally, the corporeality of an actor is the third organic dimension. All three are intermingled and, as a matter of fact, inseparable. In this totality of the decision maker, there is a minor role for mechanical occurrence. The physical processes of the body may be described in this way, but, as a totality, human action is essentially the decision maker's constant understanding of the world and himself or herself (Pihlanto 2009).

7.5 Learning

Learning is a dynamic concept that emphasizes the continually changing nature of a system (Leroy and Ramanantsoa 1997). It occurs when people detect and correct an error. An error is any mismatch between what people intend an action to produce and what actually happens when they implement that action (Argyris 1993). It is a mismatch between intentions and results. Learning also occurs when people produce a match between intentions and results for the first time.

Learning is an action concept, too (Argyris 1993). It is not simply having a new insight or a new idea. Learning occurs when people take effective action, when they detect and correct error. According to Argyris (1993), learning is intimately connected with action for three reasons:

- It is unlikely that any propositions of the if–then variety that people have stored in their heads can fully cover the richness and uniqueness of a concrete situation. There will always be a gap between their stored knowledge and the knowledge required to act effectively in a given situation. To fill the gap, learning about the new context in the new context is required.
- Even after the knowledge gap has been relatively closed, it is unlikely that the action that people design and implement will be adequate. Most contexts or situations that concern them are constantly changing. People cannot assume that other individuals or groups will react as they had thought they would when they designed their actions. There is a continual need for vigilant monitoring of their and others' actions. These processes require learning too, often performed iteratively.
- Learning is not only required to act effectively; it is also necessary to codify effective action so that it can be repeated reliably when appropriate. This means that effective actions are not only stored as rules in actors' heads; it means that their requirements are known publicly, usually in the form of formal and informal policies and routines that are rewarded by systemic cultures. Building policies, routines and culture requires learning.

Learning occurs in a manner such that, for the observer, the learned behaviour of the organism appears to be justified by the past through the incorporation of a representation of the environment that acts, modifying its present behaviour by recall. Notwithstanding this, the system itself functions in the present, and learning occurs for it as an atemporal process of transformation. An organism cannot determine in advance when to change and when not to change during its flow of experience, nor can it determine the optimal functional state that it must reach in advance. Both the advantage of any particular behaviour and the mode of behaviour itself can only be determined a posteriori as a result of the actual behaviour of the organism subservient to the maintenance of its basic circularity.

According to Maier et al. (2001), all organisms are capable of learning at different levels. The survival of each species demands that the members of the species adapt to their particular environment. This behavioural adaptation can emerge in two ways, through the evolution of the species as a whole and through adaptation by individual members. In the first case, features that have proven to be especially useful for survival are passed down from generation to generation within the species. In the second case, the organism learns and then adopts new behaviour that optimizes its survival. Of these two types of behavioural changes, evolutionary adaptation has the advantage that each individual is endowed with capabilities or skills for successful survival. Adaptation through learning is inherently risky, because the individual does not know for certain how to react to the environment during the learning process and therefore might be killed before acquiring the necessary skills for survival.

In the opinion of Allee (1997), the processes of knowledge creation can simply be thought of as learning. Learning is defined as gaining knowledge, comprehension or mastery through the experience of study. Defining the processes of acquiring or creating knowledge as learning allows individuals to distinguish between knowledge as a process and knowledge as an object. The methods for gaining knowledge can be viewed separately from those used to organize, sort, catalogue and otherwise process an existing body of knowledge. Skill-based knowledge, such as filling out a form, may only require a demonstration and a little practice. Theoretical knowledge, however, is much more learning intensive. It requires research processes, in-depth study and critical discussion with experts (Allee 1997).

When individuals decide to acquire knowledge, they require learning processes to support that effort. Their intention to learn determines the shape of their knowledge. Different intentions around learning result in different bodies of knowledge. In the same way, different types of knowledge require different learning processes.

To summarize, learning as a process consists of the transformation through experience of the behaviour of an organism in a manner that is directly or indirectly subservient to the maintenance of its basic circularity (Maturana 1970). Due to the state-determined system of the living system, this transformation is a historical process such that each mode of behaviour constitutes the basis on which a new behaviour develops, either through changes in the possible states that may arise in it as a result of an interaction or through changes in the transition rules from state to state. The organism is thus in a continuous process of becoming that is specified through an endless sequence of interactions with independent entities that select its changes of state but do not specify them.

7.6 Communication

Communicative processes are seen as informative inputs, the communication itself and informative outputs (e.g. Rescher 1996). In effect, these three stages reflect the question of symbol use conditions, the symbolic process itself and symbol use results. From this standpoint, the enterprise of communication is also to be regarded from a thoroughly processual perspective. This is only appropriate, as communication is ill served by being depicted as commerce in things as per the misleading reifications of symbols, messages and meanings. These are no more than suppositional abstractions, which are useful, perhaps, as shorthand devices for presenting considerations that themselves are of a very different status—not things of some sort but processes at work in the ideational transactions of informative thought.

This view of communication as a process of information transmission across the diversity of persons and times is characteristic of a process epistemology that sees this sort of convenience in information not as a matter of manipulating ideational objects but rather as one of conveying information with a view to its use by oneself and others within a characteristic range of processual operations. The communication of information that is indispensably necessary to the acquisition and confirmation of knowledge in a communal setting is clearly also a thoroughly processual enterprise. Moreover, the interpretation of communicated symbols as the transformation of physical signals into informative messages is itself inevitably a process, one to which the recipient also has to make an active contribution.

This circumstance has larger implications. The most fundamental and pervasive factor of reality is its instability—its being a flux. This circumstance encompasses knowledge itself. It too is in a constant state of flux. This means that reality cannot be represented and conveyed adequately by the relatively fixed and discrete mechanisms of language and its concepts. Symbolic representation of purported fact is virtually always unfaithful to reality. In other words, our descriptive formulations are not to be taken too literally. To some extent, they are generally figurative—mere likenesses. In the final analysis, our attempts to characterize the world's arrangements by linguistic means must be acknowledged as being almost inevitably metaphorical to some degree.

Therefore, our human conceptual thinking always involves some distortion of fact, some unfaithfulness to reality. Here the role of metaphors is crucial. At the bottom, linguistic communication as we know it builds on a comparatively small basis of practically geared literal language use. However, this is infinitely enriched and amplified by metaphorical flights—processes of assimilation that extend its communicative resources in every direction. The metaphorical nature of linguistic communication is yet another aspect of the many-sided, processual nature of human cognition.

From the process standpoint, then, it emerges that the purposes of an adequate understanding of knowledge acquisition and management are ill served by a problematic reification of the elements of knowledge into thought things. Human knowledge is geared towards activity. Its terms of reference are given by verbs like asserting, questioning, understanding, communicating and so forth. From the epistemological and logical points of view, those abstract items at issue in communication—meanings, data and the rest—are convenient fictions devised to describe, compare and contrast the fundamentally processual phenomena of acquiring, transmitting, storing and utilizing information. To make substance like things of cognitive and communicative matters is to commit Whitehead's fallacy of misplaced concreteness. The development, employment and communication of information in a communal setting is in all of its aspects an active and thoroughly processual enterprise.

Furthermore, Luhmann (1986, 1995a) argues that communication is the most fundamental social category, more so than an individual action. Actions need not be inherently social, whereas communications are, although this does verge on the tautological, since the social is defined as a system of communications. Social actions already presuppose communications in the sense that they rely on or raise the expectation of recognition, understanding and acceptance by others. In other words, a social action is inevitably already a communication. Nevertheless, a communication is more than simply an action. It involves and therefore includes the understanding of another party and so extends beyond the individual action to form the link necessary for social operations. A communicative act in itself leads to nothing; it is only when it generates some understanding in another that it can trigger a further communication.

It is important to understand what Luhmann means by communication, since he uses the term in a very specific sense. He stresses that it is not what we might normally mean by a communicative act, such as a statement or utterance by a particular person. Indeed, it is on a different level from people and their thoughts and actions. For Luhmann, these are not part of the social system at all but part of its environment. He characterizes a communication as an event consisting of three indissoluble elements—information, utterance and understanding—which can enable further communications to occur. Each of these elements is said to be a selection, that is, one possibility chosen (but not necessarily by a person) from many. It is the operation of the autopoietic system that defines and makes the selection. Thus, broadly speaking, information is what the message is about, utterance is the form in which it is produced together with the intention of its sender and understanding is the meaning that it generates—which can include misunderstanding—in the receiver. This means that there must be at least two parties involved in the communication, and it is this that makes communication the most basic social element for Luhmann.

All these three elements are generated or co-produced as a unity, and this event allows the possibility of further communications. This happens through a fourth selection by the receiver, the acceptance or rejection of the communication's meaning (Luhmann 1995a, p. 147). This is distinct from understanding. Any communication generates meaning, whether intended or not. The fourth selection is the link to action—does the receiver respond in some way to the communication, perhaps to question or disagree, or does the receiver fail to respond and thereby terminate the communicative sequence? It is important to stress that all aspects are distinctions made by the system itself, not by an outside observer. The system determines what, for it, is information; how it may be embodied; and how it may be interpreted. Accordingly, it draws its own distinction as to what belongs to the system and what does not—this is the closure of the autopoietic system (cf. Seidl and Becker 2005).

Thus, the actual components of communication are selections. This means that communication always entails a momentary choice of observations; that is, communication is the processing of selection. By linking communication with observation, the former comes to be regarded as a complex undertaking insofar as social systems, or rather the communications thereof, can mutually observe one another. Observing is no longer the exclusive performance of a psychic system but an abstract procedure. By means of its capacity for observation, communication can be considered as a process full of events, in which momentary decisions are made about what is being communicated. The unmarked communications are always already inherent within this process. In this way, one can always understand communication as a unity of difference.

Therefore, every communication operates as a selection; otherwise, communication cannot give meaning. It follows from this that communication cannot, socially, communicate about every matter at a present moment. This means that the social, material and temporal complexities of the world have to be reduced. Otherwise, meaningful communication is not possible (Luhmann 1984). Such a reduction is possible if the complexities of the world are selected by distinctions between what is actual and relevant and what is not. Thus, communication is coded to operate as communication.

While Luhmann thus views communications instead of actions as the elementary units of social systems, the concept of action admittedly remains necessary to ascribe certain communications to certain actors. Thus, the chain of communications can be viewed as a chain of actions. This enables social systems to communicate about their own communications and to choose their new communications, that is, to be active in an autopoietic way. Such a general theory of autopoiesis has important consequences for the epistemology of the social sciences. It draws a clear distinction between autopoiesis and observation but also acknowledges that observing systems are themselves autopoietic systems, subject to the same conditions of autopoietic self-reproduction as the systems that they are studying.

Thus, social systems proceed blindly in the autopoietic reproduction of communication, as they cannot reproduce and simultaneously observe themselves. Observations of communication are second order (von Foerster 2003). They are either further communications of the same system or autopoietic operations of other systems altogether, that is, communications of a second system or thoughts of an individual consciousness. Social systems thus oscillate between reproductive communication and retrospective observation. The unique sequence of operations maintains the evolution of systems at large. Communication as the basic element of the evolution of social systems is either reproduced in the autopoietic networks of communication or not. Although communication is indecomposable for social systems, it produces a precisely defined process (cf. Luhmann 1984, pp. 191–241, 1992).

When codified communication operates, an infinite number of codes will occur and again disappear in the course of time. Codes have to be selected, that is, in their turn are codified to achieve stability in time. Codes that reduce world complexity are objects to codes of codes. Therefore, the reduction of outer-world complexity creates inner complexity.

Moreover, whenever such internal complexity emerges, systems emerge distinguished from their environment. Hence, systems are not substantial entities or elements in arranged relations to each other, nor are they primarily characterized by their adaptability to the external world. Rather, they are distinguished from the external world and only open to selections, as information, from the external world if they operate from the base of an internal closure of their operations. Thus, systemic communication is above all self-referential and self-organizing. Self-referential systems might even become autopoietic; that is, they construct their own elements, which are their own codes, definitions and distinctions. In general, functional systems can only communicate with themselves and not with the external world. They can send information and more complex messages, but the interpretation is contingent on the internal operations of specific systems (Bakken and Hernes 2002a).

To sum up, any communication generates meaning, whether intended or not. According to Luhmann, communication is not the simple sending of a message. The event cannot be said to have occurred until the receiver has understood something, even if not what was intended. Then, the very nature of the communication remains undefined until it has been interpreted by the other. Nor can communication be understood as the transmission of something (knowledge) from inside one person's head to inside another's. The utterance is a selection, a skilled performance chosen to provoke or trigger a reaction in the receiver, but it can never determine what the reaction will be, as this is too complex a selection based on the receiver's own cognitive state. Moreover, Luhmann (1995a) conceptualizes social structures as expectations. This means that, in every situation, certain communications are expected and others are not. For example, a question about a business firm's strategy is expected to be followed by an answer to this enquiry and not by a statement about the latest football results. This means that an expectation, to a certain extent, preselects the possibilities for further communications: it makes certain communications more likely than others.

7.7 Consciousness

Psychic systems communicate on the basis of consciousness or thoughts. However, communication between psychic systems must not be understood as the transmission of knowledge from inside one person's head to inside another person's head.

Numerous experiments with human beings have shown that consciousness is composed of many dimensions. It is created by many different brain functions, yet it is a single coherent experience. For example, when the smell of a perfume evokes a pleasant or unpleasant sensation, one experiences a single, coherent mental state composed of sensory perceptions, memories and emotions. The experience is not constant and may be extremely short. Mental states are transitory, continually arising and subsiding. However, it does not seem possible to experience them without some finite span of duration. Another important observation is that the experiential state is always embodied, that is, embedded in a particular field of sensation. In fact, most mental states seem to have a dominant sensation that colours the entire experience.

All kinds of psychical–mental activities constitute the consciousness of an individual in the form of a continuous process. An object in the situation of an individual, for example a task in a production, provides the consciousness with meaningful content. A meaning emerges in the consciousness as this content becomes referred to the object located in the situation of the person in such a manner that he or she understands what the object implies. This is to say, a person can understand an object only in terms of a meaning or a group of meanings. The network of all meanings accumulated in the consciousness is called the worldview of an individual. In accordance with the autopoietic epistemology, the worldview is continuously redefined as new meanings emerge on the basis of new objects observed in one's situation.

For instance, in reading a book, an individual is so absorbed therein that he or she blocks out his or her own thoughts, which would otherwise interfere with the participation in the communicative process of reading (Maurer 2010). Consciousness is then so preoccupied with language that one's thoughts and ideas are fixated only on the communicative event. The element of communication—information, utterance and understanding—must be synthesized to facilitate further connecting communications. The thoughts of readers play no determining role, because consciousness, which selects one way or the other, is not deciding about communication. Communication itself is deciding. Nonetheless, consciousness plays a necessary role in the communicative process, which would not be possible at all without it.

Consciousness's constitutive share in communication arises from perception (Maurer 2010). Perception is a 'special competency of consciousness' (Luhmann 2000, p. 17) and a non-communicative event of consciousness. Without perception, nothing can be conveyed as having been perceived, which implies that communication depends on perception. Language can stimulate and irritate consciousness by making 'conspicuous objects of perception' (Luhmann 1993, p. 48) available. The objects of perception that can irritate consciousness are words that meet special criteria: 'They may not present any similarity to other perceivable objects (sounds, imaged, etc.); for that would cause them to continually seep back into the world of perception and disappear therein' (Luhmann 1994, p. 48). Words must be specifically constituted to avoid being reduced back into the world of perception. This also means that their characteristics must be constantly preserved so that they are always utilizable. Only through the regularities implied by words is it possible for consciousness to irritate communication in such a way that generates communication that is more complex and more differentiated than is possible through gestures, for example. Only once these preconditions are met can one understand how linguistic communication can attract the attention of consciousness: 'Consciousness can therefore hardly withdraw itself from a communication in progress. At most, it can, while listening, entertain extravagances or attempt to irritate communication with its own contributions' (Luhmann 1994, p. 48).

Everything in this process occurs in terms of understanding, which means that a person knows, feels, believes in and dreams about the phenomena and objects located in his or her situation in terms of their being something. Understanding is complete only after a meaning is generated. Meanings are components from which the world, as people experience it, is constructed. In the consciousness, continuous restructuring of meanings occurs as a person actively acquires or passively gains knowledge from the situation, for example observes and learns new things. Meanings are often forgotten, fading into the unconsciousness and perhaps retrieved into the consciousness anew. An important condition is that all aspects of knowledge and skills are stored in the worldview of an individual in terms of different kinds of meanings (cf. Pihlanto 2005).

What an individual in a system brings to the knowledge creation situation has an important influence on what he or she can learn from another individual. This means that an individual's personal worldview profoundly influences the way in which he or she experiences the situation at hand. '... although it is the individual who learns, this individual is one who has a language, a culture, and a history' (Usher 1989, p. 32). Thus, an individual's personal worldview affects, for example, how he or she commits to the task at hand and what he or she can understand in the first place about the knowledge communicated. People always learn in relation to their worldviews or what they have learned before (cf. autopoietic epistemology in Chap. 4.1.3).

Psychic systems are particularly important for social systems. Like social systems, psychic systems are meaning-constituted systems. However, in contrast to social systems, the meaning events do not materialize as communications but as thoughts. In other words, psychic systems reproduce themselves on the basis of consciousness; that is, only thoughts can produce thoughts. As operatively closed systems, psychic and social systems constitute environments for each other. That is, thoughts cannot become communications and communications cannot become thoughts. Mutual influences are restricted to the structural level. There merely exists a relation of structural coupling, that is, both types of systems are structurally adapted to each other in a way that allows for mutual perturbations. Luhmann calls the specific structural coupling of social and psychic systems interpenetration. Then, Luhmann speaks of interpenetration if an autopoietic system presupposes the complex achievements of the autopoiesis of another system and can treat them like parts of its own system.

To sum up, an individual understands an object only in terms of meaning. This means that an individual's existing worldview determines how a piece of data (i.e. perturbation) is interpreted. The information process may be influenced by the individual's position within the system, previous experiences and other people and the environment. To establish uniformity of shared interpretation, there needs to be uniformity in worldviews among the people of a system. This is easier when new triggering perturbations are framed in a consistent and familiar manner. If a new perturbation is framed in a different manner around different people within a system, it is likely that there will be a diversity of shared understandings of the perturbation.

7.8 Relation Between Social and Psychic Systems

The simultaneous—but separate—autopoiesis of psychic systems is constitutive of the autopoiesis of the social system. Without psychic systems, social systems are impossible. Every communicative event presupposes parallel events in psychic systems. For the perception of utterances, the social system depends at any rate on the psychic system; the social system cannot hear spoken words or read letters. Moreover, psychic systems serve as a memory, as they can remember communicative events beyond their momentary point of existence. Because of their structural coupling, social systems can expect their communications to cause perturbations in the psychic systems and to receive perturbations from the psychic systems when necessary. For example, they can count on psychic systems to trigger further communications after every communication. Although psychic systems trigger communication processes and vice versa, the processes of the psychic system and the social system do not overlap in any way (cf. Seidl and Becker 2005).

According to Seidl and Becker (2005), the most important evolutionary achievement for the coupling of social and psychic systems is language. However, this does not mean that communication is possible only with language. Language ensures that psychic systems are perturbed through the communication processes. That is, articulated speech normally disturbs people who are not involved in the communication more than mere noise does. Thus, psychic processes are synchronized with communication processes and in that way they know when to contribute perturbations to the communication process to make the reproduction of the social system possible.

In the opinion of Seidl and Becker (2005), although Luhmann's strict distinction between social and psychic systems runs counter to our everyday beliefs and almost all social and psychological theories, it has one important theoretical advantage. It allows for a concept of the social realm, which is clearly distinguished from the psychological. Consequently, social and psychic phenomena can be analysed in their own right. This does not lead to the marginalization of psychic systems from social systems, as criticism has often suggested. On the contrary, through this differentiation, it can be clearly shown that the two types of systems depend on each other.

To summarize, as stated earlier, autopoiesis is a concept that differentiates the living from the non-living. An autopoietic system is defined as a network of interrelated component-producing processes such that the components in interaction generate the same network that produced them. Although Maturana and Varela consider the concept to be applicable only in biology, and not in the social sciences, an interesting theory transfer is made by Luhmann (1988). He defends the quite novel thesis that, while social systems are self-organizing and self-reproducing systems, they consist not of individuals—or roles or even acts, as commonly conceptualized—but exclusively of communications. When generalizing the usages of autopoiesis, developed while studying biological systems, to make it also applicable to social systems, the biology-based theory of autopoiesis should therefore be expanded into a more general theory of self-referential autopoietic systems. Social and psychic systems are based on another type of autopoietic system than living systems, namely on communication and consciousness, respectively, as modes of meaning-based reproduction.

7.9 Summary

The notion of a social autopoietic system is based on the ideas of Niklas Luhmann. This chapter has explored some claims and suggestions regarding this type of system.

Social autopoietic systems are realized in a domain of communications. In other words, the constituent elements of social autopoietic systems are communications. Thus, a social autopoietic system is understood as a network of communications.

Apart from living systems, there are two additional types of autopoietic systems: social systems and psychic systems. While living systems reproduce themselves on the basis of life, social systems reproduce themselves on the basis of communication and psychic systems on the basis of consciousness. A social autopoietic system is in a continual process of recreation of itself anew as communications generate further communications and thereby internal structure.

Events mark the selection of some alternatives over others. If an unfortunate decision is made, correcting it will not eliminate that decision, but its correction will enter the process as a new event. Articulation serves to bring meanings to life, which in turn provide acts with meaning. Meanings are constituted by accumulations of events over time; hence, they may be seen as event-objects. Events provide meanings with historicity and direction.

In the autopoiesis theory, self-awareness is viewed as being tied closely to language, and the understanding of language is approached through a careful analysis of communication. Language and languaging are useful interpretative lenses through which to understand different knowledge creation processes. The central element around which any viable concept of system has to be built is a decision, yet the concept of a decision generally seems to be even less clear than that of a system. Usually a decision is defined as a choice.

The processes for knowledge creation can be thought of simply as learning. Learning is defined as gaining knowledge, comprehension or mastery through experience of study. Communication as an event consists of three indissoluble elements—information, utterance and understanding—which can enable further communications to occur. Each of these elements is a selection.

Chapter 8
Knowledge Creation

From a mainstream perspective, knowledge creation is basically a process of transmission between individuals in which data are converted into information through the medium of knowledge, which may be explicit but, far more importantly, may be tacit. The transmission of knowledge between people is a process of conversion between tacit and explicit forms based on mimicry in tacit–tacit transfers, group dialogue and discussion in metaphorical and analogical language in tacit–explicit transfers, formalization and codification in explicit–explicit transfers and internalization in explicit–tacit transfers. Knowledge is understood to move in this way through the interplay of individual and group/systemic/social levels (Nonaka and Takeuchi 1995; Stacey 2001).

Knowledge and knowledge creation are both constructs in that they are abstract theoretical variables that are invented to explain phenomena. Among the key concerns about construct measurement is that of validity. To be judged as adequate, a measure must reflect the theoretical content domain of the construct (content validity). To assess content validity, the construct is theoretically defined, incorporating a clear specification of the total content universe that is relevant, and a representative measure is drawn from the content domain, which is operationalized to reflect the meaning of each chosen dimension. Therefore, the initial step towards establishing the content validity of knowledge creation measures is to review the theoretical definitions of knowledge creation used in current research and to undertake an assessment of their success in clearly specifying relevant content.

Boisot (1998) adopts the mainstream definition of knowledge as the capacity to act and states that it is built on information that is extracted from data. Effective action is dependent on representations that connect with the real world, and those representations are formed into mental models to make sense of the world. This means that knowledge creation is a process of generating insight by extracting information from data and the application of knowledge is the testing of these insights. Knowledge constitutes an asset that yields a stream of useful services. Its distinctive feature is that it can be shared with others and retained at the same time. In

K. U. Koskinen, R. Breite, *Uninterrupted Knowledge Creation*, SpringerBriefs in Business, https://doi.org/10.1007/978-3-030-57303-4_8

the opinion of Boisot (1998), there is no questioning of, or departure from, mainstream thinking here.

One of Boisot's principal concerns is with the conditions in which knowledge will flow, be shared with others, and those in which it will not (Stacey 2001). In some circumstances, knowledge diffuses naturally, while in others, it does not, and understanding these circumstances is essential to mastering knowledge. Boisot claims that knowledge flows between people, that is, it is fluid, when it is context-free, codified, abstract and stripped of unnecessary data. When knowledge is data rich, qualitative, ambiguous and context-dependent, then it is viscous and hardly flows at all. Over time, as one comes to understand something, viscous knowledge becomes more fluid, enhanced by personal experience and idiosyncratic interpretation as it is embedded in individual minds. However, people find this fluid knowledge hard to share. Codification is then required to diffuse the knowledge in a system. However, this makes it more accessible to competitors and thus less valuable.

8.1 Knowledge Creation Process

Nonaka and Takeuchi (1995) see ongoing knowledge creation as the source of continuous innovation. When systems innovate, they do not simply process information, from the outside in, to solve existing problems and adapt to a changing environment. They actually create new knowledge and information, from the inside out, to redefine both problems and solutions and, in the process, to recreate their environment.

Thus, they consider knowledge as a dynamic human process of justifying personal belief in the truth. This understanding emphasizes that knowledge is essentially related to human action. As a fundamental basis for the theory of systemic knowledge creation, they focus attention on the active, subjective nature of knowledge represented by such terms as commitment and belief, which are deeply rooted in individuals' value systems. This means that the basic argument is that knowledge creation is a synthesizing process through which a system interacts with individuals and the environment to transcend the emerging contradictions that the system faces.

Furthermore, according to Nonaka and Takeuchi (1995), the knowledge-creating process is a continuous, self-transcending process. As knowledge is created between individuals or between individuals and the environment, individuals transcend the boundary between self and others. Hence, following Nonaka and Takeuchi (1995), there are four types of knowledge-creating processes (the SECI model, Fig. 8.1).

Fig. 8.1 SECI model (Source: Nonaka and Takeuchi 1995)

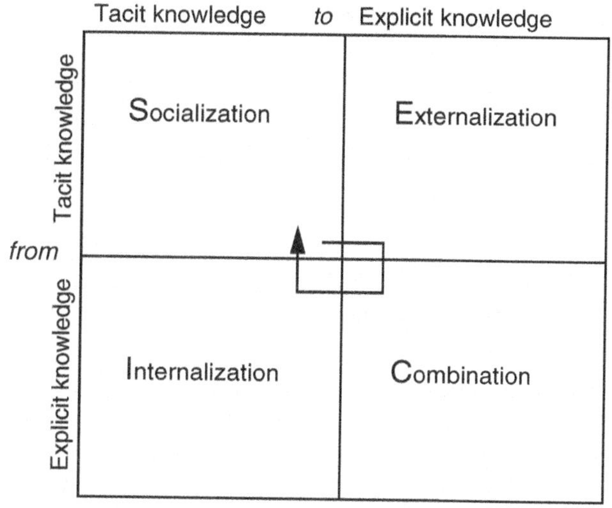

8.1.1 Socialization

This process focuses on tacit-to-tacit knowledge linking. Tacit knowledge extends beyond the boundary, and new knowledge is created by using the process of interactions, observing, discussing, analysing, spending time together or living in the same environment. Socialization is also known as converting new knowledge through shared experiences. Systems gain new knowledge from outside their boundary too, like interacting with customers, suppliers and stage holders. This occurs in traditional environments in which a son learns the technique of woodcraft from his father by working with him (rather than from reading from books or manuals).

8.1.2 Externalization

This process focuses on tacit-to-explicit knowledge linking. It helps in creating new knowledge as tacit knowledge emerges from its boundary and becomes collective group knowledge. Through this process, we can say that knowledge is crystallized. The process of externalization is often driven by metaphors of analogy and models. Quality circles are formed in manufacturing sectors in which workmen put their learning and experience into improving or solving process-related problems.

8.1.3 Combination

Combination is a process whereby knowledge transforms explicit knowledge into explicit knowledge. For example, the creative use of a database to obtain a business report, sorting, adding and categorizing are some examples of combination processes.

8.1.4 Internalization

Through internalization, explicit knowledge is created by using and sharing tacit knowledge across the system. When this tacit knowledge is read or practised by individuals, it broadens the learning creation in a system. The system tries to innovate or learn when this new knowledge is shared in the socialization process. Systems provide training programmes for their employees at different stages of their work, for example within a business company. By reading these training manuals and documents, employees internalize the tacit knowledge and try to create new knowledge after the internalization process.

Thus, according to the SECI model, knowledge is continuously converted and created as users practise, collaborate, interact and learn. Therefore, the process should be seen as a continuous, dynamic swirl of knowledge rather than a static model. It is basically a visual representation of the overlapping, continuous processes that take place—or should take place—in a system.

8.2 Two Major Knowledge Flows

A system must continuously replenish its stock of knowledge necessary to use different resources effectively. This is because a system's operations are determined by the specific activities or systemic routines that it can perform in using its resources (Nelson and Winter 1982). This means that knowledge flows have a crucial role in the reproduction of a system's knowledge structure (Tuomi 1996), hence the idea that maintaining a system requires processes of sensing (a condition for interactive openness) and memory (a feature of self-referentiality), each of which constitutes a major knowledge flow (e.g. Maula 2006). This means that an autopoietic system incorporates two major knowledge flows: sensing and memory. Sensing means, in practice, that the system interacts, co-evolves and coordinates its activities with its changing environment. For example, a system creates new knowledge by using various kinds of boundary elements, such as roles and functions, through which it interacts reciprocally with its environment. Memory, in turn, means that the system has access to its own accumulated knowledge. The system is, therefore, internally closed in the sense that it utilizes its existing knowledge resources and may thereby operate efficiently. Sensing and memory help the system to make distinctions that then become embedded in its internal knowledge structure.

8.2.1 Sensing as a Source of a Major Knowledge Flow

To survive, adapt, learn and renew itself, a system needs the ability to co-evolve reciprocally with its environment. Boundary elements influence a system's learning and renewal capability by enabling two kinds of sensing activities:

- Exposure or awareness of the system to triggers—perturbations in its environment that elicit compensation reactions
- Interactive processes and communication with other entities.

In other words, these two activities enable a system to maintain openness. In this way, autopoietic boundary elements function as connecting and absorbing mechanisms rather than as separating elements.

Sensing means that a system interacts with its environment by being aware of, and compensating for, perturbations, by improving its knowledge and by changing internally. In other words, a system interacts with its environment through structural coupling, that is, through recurrent interactions, each of which triggers changes in the system. However, it is crucial to realize that the environment only triggers changes; it does not specify or direct them (cf. Maturana and Varela 1980).

Indeed, structural coupling establishes a clear difference between the ways in which living systems (e.g. business firms) and non-living systems (e.g. machines) interact with their environments. As mentioned earlier, Bateson (1972, 1979) points out that kicking a stone and kicking a dog are two very different stories.

To conclude, as a system responds to environmental influences with internal structural changes, these changes will in turn alter its future behaviour. This means that a structurally coupled system is a learning system. As long as a system remains functioning, it will couple structurally to its environment. Its continual structural changes in response to the environment—and consequently its continuing adaptation, learning and development—are key characteristics of the behaviour of systems.

8.2.2 Memory as a Source of a Major Knowledge Flow

Psychological research makes a distinction between learning and memory (e.g. Postman 1976). Learning has more to do with acquisition, whereas memory concerns the retention of whatever is acquired. In reality, however, separating these two processes is difficult, because they are tightly interconnected. What we already have in our memory affects what we learn, and what we learn affects our memory (Kim 1998) (cf. autopoietic epistemology in Chap. 4.1.3). The concept of memory is commonly understood to be analogous to a storage device, into which everything that we perceive and experience is filed away.

While sensing helps the system to coordinate its functions within the changing environment and to create new knowledge, memory maintains the system's daily functioning and utilization of accumulated knowledge. According to Maula (2006), this self-referentiality means that:

- The accumulated knowledge affects the system's mode of operation
- The mode of operation affects the creation and acquisition of new knowledge.

According to Hofstadter (1979), different systemic levels collapse, which makes it possible to understand the phenomenon of self-reference. These levels should be interpreted at the same time in terms of being separated and tangled, hierarchized and non-hierarchized. In the opinion of Hofstadter, this helps us to understand strange loops, which threaten the stability of the hierarchy and may even lead to its destruction. Every objective takes the place of another in a process of oscillation that cannot be stopped (cf. Bakken and Hernes 2002a). Hence, self-referentiality can be a resource, but it can also be a constraint, depending on the implementation.

Thus, memory provides access to accumulated experience and knowledge. Knowledge is stored in the system's internal knowledge structure—such as a system's shared culture, strategies, rules and practices. It can be stored in the competence of the system and the expertise of its individuals. Knowledge can also be stored in explicit form in databases.

Because there is a variety of knowledge, and the needs of systems vary widely, it is important that knowledge management practices support the specific needs of the system. For example, a small system with a high degree of tacit knowledge may reduce the need to accumulate explicit knowledge. On the other hand, a large system that operates globally often shares and reuses explicit knowledge with the help of information technology. This creates opportunities to save time in routine parts of an assignment and to focus on those parts that require innovation and expertise.

In sum, systems utilize two processes, one that provides new knowledge for the system and coordinates it with the environment (sensing) and another that provides access to existing knowledge and increases effectiveness (memory). The processes are interconnected, for example, through information and communication systems. This means that a continual coordination of these flows is necessary so that new knowledge becomes a part of the existing knowledge structure and the existing knowledge structure helps to find, create and evaluate new knowledge (cf. Maula 2006).

8.2.3 Uninterrupted Knowledge Creation

A common charge against the autopoietic perspective is that it does not explain change, hence knowledge creation. Autopoiesis means self-production, and therefore the term lends an impression of systems engaging exclusively in maintaining, unchanged, their basic features. The charge is therefore understandable given the assumption that the only way for systems to maintain themselves is to close themselves off from other systems, hence the impression that nothing can trickle in from the external environment and that novelty cannot be created (Hernes 2008).

However, according to Hernes (2008), two things can be said about stability and change in relation to an autopoietic perspective. First, developing stability should be seen as a change in itself. One of the great achievements of systems is their capacity to transform unorganized complexity into organized complexity while building the external complexity into their internal operations. This is how routines are developed to rationalize operations, enabling the system to meet the demands from its external environment. Thus, an important achievement of systems is their ability to transform complexity into simple structures, leaving them free to handle other areas of complexity through non-routinized behaviour. When it comes to complexity, most process approaches would dictate a pragmatic view of the degree of complexity that systems can handle. In Luhmann's thinking, this means that, beyond a certain level of complexity, autopoiesis is not possible. A subject of study then becomes the transition of one level of complexity to another and how new systems are formed in response to increased complexity. In relation to systems, the point is highly relevant, notably in relation to understanding how systems differentiate internally to cope with simultaneously competing institutional pressures from their environments.

Second, according to Hernes (2008), continuity offers an opportunity for change. According to the principle of autopoiesis, systems uphold themselves through interaction with their own states. Without reproduction, a system breaks down. On the other hand, if there is only reproduction of the existing features of the system, the system cannot change and will remain essentially identical over time. The dilemma is resolved by conceptualizing the relationship between process and structure. A process, consisting of successive events, offers occasions for change as well as continuity (Luhmann 1995a, p. 347). Change may happen in a number of ways (such as through accidents or unintended consequences), but it will take hold only insofar as it can be understood by the system, that is, interpreted through the codes of communication that are appropriate to the system in question. Structure presupposes self-maintenance that is sufficiently stable to enable meaning to be made of opportunities for change, thus enabling choices to be made against a horizon of recognizable possibilities. The discussion of continuity and change in relation to structure and process, respectively, offers a framework that greatly benefits the study of systemic change, hence knowledge creation.

Luhmann's general concept of autopoiesis radicalizes the temporal aspect of autopoiesis. While Maturana and Varela conceptualize the elements of biological systems as relatively stable chemical molecules, which have to be replaced from time to time, Luhmann conceptualizes those elements as momentary events without duration that vanish as soon as they come into being; they are momentary and immediately pass away. Thus, because the components (i.e. events) of the system have no duration, the system is urged to produce new components constantly.

Furthermore, according to the traditional stable process problem (i.e. the macro–micro problem), the system's structure (i.e. stable) and production (i.e. process) cannot interact (e.g. Bakken and Hernes 2002a). Conceptually, they remain distinctly different entities, and the differences stem from epistemologically different theoretical projects. However, the complexities of systems demand that we are able to analyse them at different levels—specifically the system's structure and production levels—and that we are able to relate processes at different levels to one another.

Thus, on the basis of autopoiesis theory, the interaction between process (e.g. production) and stability (e.g. structure) takes place through the operation referred to as recursivity (Luhmann 1995a). Recursivity is that which permits the reproduction of interactions over time. Having a recursive view of a system implies dealing with the question of how this system persists and develops (e.g. Koskinen 2013). For example, recursivity takes place when the knowledge needed by the system is offset against the system's present knowledge, which again enables new knowledge to occur. This means that a system's knowledge is developed by production, which, in turn, influences future production. According to Giddens (1984), recursivity occurs in the field of tension between structure and actions, that is, a system's structure and production, respectively. Hence, a system's structure and production become mutual media for one another in recursive processes.

In addition, as stated earlier, continuity offers opportunity for change. According to the principle of autopoiesis, systems uphold themselves through interaction with their own states. Without reproduction, a system breaks down. On the other hand, if there is only reproduction of the existing features of the system, the system cannot change and will remain essentially identical over time. On the basis described above (in the chapter Change and Becoming), the dilemma is, according to Hernes (2008), resolved by conceptualizing the relationship between process and structure. A process, consisting of successive events, offers occasions for change as well as continuity.

Although production is recursive, people prefer to think that there is a level beyond the production (e.g. a system's structure) that provides a context for production. This level is not a level unaffected by production in the system. Instead, this other level is both produced by production and influences production in turn. Recursivity refers principally to the interaction between production and its context.

For example, a system's knowledge structure (i.e. system identity, memory and culture) is created through production, that is, it forms the context within which production takes place. Although the system's knowledge structure was created in the past, it is formative for future production. It is, therefore, impossible to understand the future without understanding the past, as the past is written into the future. Seen in this way, a system's knowledge structure may constitute constraints, partly because it is created in the past and partly because it puts limits on potential production.

Thus, when a system's (e.g. a business firm's) new production is planned, a repertoire of possibilities is open to the people involved. This repertoire is shaped by previously completed production. Whether it is expected or unanticipated, it serves to inform new production. Hence, the idea of recursivity represents explanatory potential for relationships between the system's knowledge structure (i.e. system identity, memory and culture) and production within the system in ways that are not possible with singular epistemologies. This takes place by considering these systems as wavering between change and no change and understanding relationships among the past, the present and the future so that new insights may be gained (cf. Bakken and Hernes 2002b).

Therefore, a system (e.g. a business firm) serves to bind different constituents (e.g. pieces of knowledge forming a system identity, memory and culture, i.e. knowledge structure) over time. This means that it is inconceivable that the system can exist without such bonds. The system presupposes interaction around these constituents and provides the essential stabilization of expectations among those who take part in these activities. It is equally inconceivable that a system should exist without production. In the absence of production, there is nothing to inform systems, so they would not be able to reproduce themselves. Thus, a system's knowledge structure and production are mutual media for one another in recursive processes. Thus, having a recursive view of a system implies dealing with the question of how this system persists and develops.

Furthermore, recursivity is pivotal to the idea of autopoiesis. Autopoietic systems (e.g. a business firm), in contrast to allopoietic systems, exist through their own production and reproduction. Rather than analysing systems as entities existing on an input–output basis in relation to their environments, the emphasis is on understanding how systems reproduce themselves (Bakken and Hernes 2002a).

Thus, the idea of recursivity represents explanatory potential for bringing new light to the relationships between change and stability, hence uninterrupted knowledge production. It is by considering a system as wavering between change (i.e. production) and no change (system's memory) and understanding relationships between past, present and future that new insights may be gained.

All in all, on the basis of the above, it is proposed that a social autopoietic system's recursive production is the means by which uninterrupted knowledge creation takes place.

8.2.4 Summary

Knowledge and knowledge creation are both constructs in that they are abstract theoretical variables that are invented to explain phenomena. Among the key concerns about construct measurement is that of validity.

Knowledge is a dynamic human process of justifying personal belief regarding the truth. This understanding emphasizes that knowledge is essentially related to human action. Knowledge flows have a crucial role in the reproduction of an autopoietic system's knowledge structure. An autopoietic system incorporates two major knowledge flows: sensing and memory.

On the basis of autopoiesis theory, the interaction between process (e.g. production) and stability (e.g. structure) takes place through the operation referred to as recursivity. Recursivity is that which permits the reproduction of interactions over time.

Epilogue

Most fundamentally this book is inspired by the systemic view, Nicholas Rescher's and his followers' process views, Matura's and Varela's autopoiesis theory and Niklas Luhmann's social autopoietic system perspective. The systemic view provides a basic approach through which we may advance our understanding of knowledge creation. However, numerous books and papers dealing with knowledge creation do not take into account a system's emergent properties, which may cause essential and surprising results when different pieces of knowledge are created. Therefore, the existence of a system can only be understood through the systemic view. A radical step within the systemic view was taken in the 1970s with the development of the concept of self-referential systems. One of the most important contributions to this new phase of systems theory was the theory of autopoiesis.

However, a common charge against the autopoietic approach is that it does not explain the uninterrupted change of a system. A solution offered by this book is the idea of recursivity, which has explanatory potential for bringing new light to the relationships between change and stability. In other words, the discussion of continuity and change offers an explanation that benefits the study of systemic change, that is, uninterrupted knowledge creation.

In process philosophy, the world is an organic web of interrelated processes or series or events in which everything exists in relation. The idea of business firms (system) as goal attaining based on normative–rational models should be abandoned. Instead, we should work from empirical descriptions of how business firms operate their own production and reproduction. Seen this way, business firms are unpredictable historical systems that always operate in the present time, which they have brought forward themselves through self-referencing. All is process!

In the discourse on systems, it has been traditional to treat the nature of systems as given, and the focus has been on behaviour within a taken-for-granted context. However, recent developments in the field have started to redress this imbalance by placing the concept of systems itself in question and, in particular, by focusing on systems as processes rather than on systems as entities. It is already clear that, once

K. U. Koskinen, R. Breite, *Uninterrupted Knowledge Creation*, SpringerBriefs in Business, https://doi.org/10.1007/978-3-030-57303-4

this shift has been made, even what constitutes a system is to be seen in radically different ways. Nevertheless, the penetration of such approaches into micro-level systemic activity has, so far, been somewhat limited. A concept like processual knowledge creation cannot be dismissed because it is so prevalent.

All in all, humans are today in the midst of great change. It is a shift towards a knowledge-based economy, in which knowledge and its creation are the most important resource. This means that knowledge and its creation are the intellectual wealth of systems like business companies.

References

Åkeström Andersen N (2003) The undecidability of decision. In: Bakken T, Hernes T (eds) Autopoietic organization theory: drawing on Niklas Luhmann's social systems perspective. Abstract, Oslo, pp 235–258

Alberts B, Bray D, Lewis J, Raff M, Roberts K, Watson JD (1989) Molecular biology of the cell. Garland, New York

Aldrich H (1999) Organizations evolving. Sage, London

Allee V (1997) The knowledge evolution. Butterworth-Heinemann, Boston

Anderson V, Johnson L (1997) Systems thinking basics: from concepts to causal loops. Pegasus Communications, Cambridge

Andrew AM (1989) Self-organizing systems. Gordon and Breach Science Publishers, New York

Argyris C (1993) Knowledge for action: a guide to overcoming barriers to organizational change. Jossey-Bass Publishers, San Francisco

Arnoldi J (2006) Autopoiesis. Theory Cult Soc 23:2–3

Ashby WE (1968) Principles of the self-organizing system. In: Buckley W (ed) Modern systems research for the Behavioural scientist. Aldine Publishing Company, Chicago, pp 108–118

Astley WG, Zammuto RF (1992) Organization science, managers, and language games. Organ Sci 3:443–461

Baecker D (1996) Kybernetik Zweiter Ordnung (second order cybernetics). In: von Foerster H (ed) Wissen und Gewissen: Versuch einer Brucke (Knowledge and conscience: trial of a bridge). Suhrkamp, Frankfurt am Main, pp 17–23

Bakken T, Hernes T (2002a) Autopoietic organization theory: drawing on Niklas Luhmann's social systems perspective. Abstrakt, Liber, Copenhagen Business School, Copenhagen

Bakken T, Hernes T (2002b) The macro-micro problem in organization theory: Luhmann's autopoiesis as a way of handling recursivity. In: Bakken T, Hernes T (eds) Autopoietic organization theory: drawing on Niklas Luhmann's social systems perspective. Abstrakt, Oslo, pp 53–74

Bateson G (1972) Steps to an ecology of mind: collected essays in anthropology, psychiatry evolution, and epistemology. University of Chicago Press, Chicago

Bateson G (1979) Mind and nature: a necessary Unity (advances in systems theory, complexity, and the human sciences). Hampton Press, New York

Batterman R (2001) The devil in the details: asymptotic reasoning in explanation, reduction, and emergence. Oxford University Press, Oxford

K. U. Koskinen, R. Breite, *Uninterrupted Knowledge Creation*, SpringerBriefs in Business, https://doi.org/10.1007/978-3-030-57303-4

Baum JAC, Singh JV (1994a) Organizational hierarchies and evolutionary processes: some reflections on a theory of organizational evolution. In: Baum JAC, Singh JV (eds) Evolutionary dynamics of organizations. Oxford University Press, Oxford, pp 3–30
Baum JAC, Singh JV (1994b) Organization–environment coevolution. In: The evolutionary dynamics of organization. Oxford University Press, Oxford, pp 379–402
Bednarz J (1988) Autopoiesis: the organizational closure of social systems. Syst Res 5:57–64
Bell S, Morse S (1999) Sustainability indicators: measuring the immeasurable. Earthscan, London
Berger P, Luckmann T (1966) The social construction of reality: a treatise in the sociology of knowledge. Doubleday, Garden City
Bergson H (1922) Creative evolution. Macmillan, London
Bergson H (1946) The creative mind. Carol Publishing Group, New York
Bickerton D (1993) Language and species. University of Chicago Press, Chicago
Black M (1962) Models and metaphors. Cornell University Press, Ithaca
Blaschke S (2008) Structures and dynamics of autopoietic organizations: theory and simulation. Dissertation Universität Marburg, Wiesbaden
Boisot M (1998) Knowledge assets: securing competitive advantage in the knowledge economy. Oxford University Press, Oxford
Brown JS, Duguid P (1991) Organizational learning and communities-of-practice: toward a unified view of working, learning, and innovation. Organ Sci 2(1):40–57
Brown JS, Gray ES (1995) The people are the company. Fast Company 1(1):78–82
Bruner JS, Anglin JM (1973) Beyond the information given. Norton, New York
Capra F (1996) A new understanding of living systems: the web of life. Anchor Books, New York
Checkland P (1997) Systems thinking, systems practice. Wiley, Chichester
Chia R (1994) The concept of decision: a deconstructive analysis. J Manag Stud 31:781–806
Chia R (2000) Time, duration and simultaneity: rethinking process and change in organizational analysis. Paper presented at the Academy of Management Conference, Toronto, August 2000
Chomsky N (1968) Language and mind. Harcourt Brace Jovanovich, New York
Choo C (1988) The knowing organization. Oxford University Press, New York
Cohen MD (1991) Individual learning and organizational routine: emerging connections. Organ Sci 2(1):135–139
Cooper R (1986) Organization/disorganization. Soc Sci Inf 25:299–335
Cooper JW (2006) Panentheism – the other god of the philosophers: from Plato to the present. Baker Academic, Grand Rapids, p 176
Currie G, Brown AD (2003) A narratological approach to understanding processes of organizing in a UK hospital. Hum Relat 56(5):563–586
Czarniawska B (1997) Narrating the organization: dramas of institutional identity. University of Chicago Press, Chicago
Darwin C (1859) On the origin of species. John Murray, London
Dawson P (1994) Organizational change: a processual approach. Paul Chapman, London
Deleuze G (2004) Difference and repetition. Continuum, London
Denning S (2001) The springboard: how storytelling ignites action in knowledge-era organizations. Butterworth Heinemann, Boston
Di Paolo EA (2009) Overcoming autopoiesis: an enactive detour on the way from life to society. In: Magalhaes E, Sanchez R (eds) Autopoiesis in organization theory and practice. Emerald Group Publishing Limited, Bingley
Dodgson M (1993) Organizational learning: a review of some literatures. Organ Stud 4(3):375–394
Drucker PF (1993) Post-capitalist society. Harper Collins, New York
Easton G, Araujo L (1993) Language, metaphors and networks. Adv Int Mark 5:67–85
Eisenberg EM (1984) Ambiguity as a strategy in organizational communication. Commun Monogr 51:227–242
Ekinge R, Lennartsson B (2000) Organizational knowledge as a basis for the management of development projects. In: Accepted for discovering connections: a renaissance through systems learning conference. Michigan, Dearborn

Elias N (1978) What is sociology? Hutchinson, London
Emirbayer M (1997) Manifesto for a relational sociology. Am J Sociol 103(2):281–317
Fagot-Largeau A (2006) Ontology of becoming. College of France, France
Feldman MS (2000) Organizational routines as a source of continuous change. Organ Sci 11 (6):611–629
Flood RL (1990) Liberating systems theory. Plenum Press, New York
Foss NJ, Knudsen C, Montgomery C (1995) An exploration of common ground: integrating evolutionary and strategic theories of the firm. In: Montgomery CA (ed) Resource-based and evolutionary theories of the firm: toward a synthesis. Kluwer, London, pp 1–18
Freifelder D (1983) Molecular biology. Science Books International, Boston
Gaines BR (1979) General systems research: quo vadis? Gen Sys Yearb 24:1–9
Gherardi S, Nicolini D (2001) The sociological foundations of organizational learning. In: Giddens A (ed) (1984). The constitution of society: outline of the theory of structuration. Polity Press, Cambridge
Giddens A (1984) The constitutions of society. Polity Press, Cambridge
Giddens A (1990) The consequences of modernity. Polity Press, Cambridge
Gilmore R (1981) Catastrophe theory for scientists and engineers. Wiley, New York
Goldman AI (1986) Philosophical application of cognitive science. Harvard University Press, Cambridge
Golinelli GM (2009) L'approccio sistemico vitale: nuovi orizzonti di ricerca per il governo dell'impresa. Sinergie 80:IX–XXII
Haines SG (1998) The manager's pocket guide to systems thinking and learning. HRD Press, Amherst
Hall WP (2005) Biological nature of knowledge in the learning organization. Learn Organ 12 (2):69–188
Hall AD, Fagen RE (1968) Definition of system. In: Buckley W (ed) Modern systems research for the Behavioural scientist. Aldine Publishing Company, Chicago, pp 81–92
Haraway DJ (1991) Simians, cyborgs, and women: the reinvention of nature. Free Association Books, London
Harnden R (1990) The languaging of models: the understanding and communication of models with particular reference to Stafford Beer's cybernetic model of organization structure. Syst Pract 3 (3):289–302
Harrison PR (1995) Niklas Luhmann and the theory of social systems. In: Roberts D (ed) Reconstructing theory: Gadamer, Habermas, Luhmann. Melbourne University Press, Victoria, pp 65–90
Hatch J (2010) Defining organizational knowledge: turning individual knowledge into organizational intellectual capital. http://knol.google.com/k/defining-organizational-knowledge
Hernes T (2008) Understanding organization as process: theory for tangled world. Routledge, London
Hernes T (2014) A process theory of organization. Oxford University Press, Oxford
Hernes T, Weik E (2007) Organization as process: drawing a line between endogenous and exogenous views. Scand J Manag 23(3):251–264
Hofstadter DR (1979) Goedel, escher, bach: an eternal golden braid. Basic Books, New York
Horrocks A, Callahan JL (2006) The role of emotion and narrative in the reciprocal construction of identity. Hum Resour Dev Int 9(1):69–83
Husserl E (1948) Erfahrung und Urteil: Untersuchungen zur Genealogie der Logik (Experience and opinion: studies on genealogy of logic). Claassen and Goverts, Hamburg
Husserl E (1950) Ideen zu einer Phänomenologic und phänomenologishen Philosophie (thoughts about a phenomenology and phenomenological philosophy). Husserliana, 3. Nifoff, The Hague
Hustwit JR (2007) Process philosophy. In internet encyclopedia of philosophy. In pursuit of a holistic worldview. https://www.iep.utm.edu/. Accessed 24 Aug 2019
Ingold T (1986) Evolution and social life. Cambridge University Press, Cambridge
Jackson MC (2000) System approaches to management. Kluwer/Plenum, New York

Jackson MC (2003) Systems thinking: creative holism for managers. Wiley, Chichester
Jackson MC (2009) Systems thinking: creative holism for managers. Wiley, Chichester
James W (1996) A pluralistic universe. University of Nebraska Press, Lincoln
Jantsch E (1980) The self-organizing universe: scientific and human implication of the emerging paradigm on evolution. Pergamon Press, Oxford
Kim DH (1998) The link between individual and organizational learning. In: Klein DA (ed) The strategic management of intellectual capital. Butterworth-Heinemann, Boston, pp 41–62
Kim DH (1999) Introduction to systems thinking. Pegasus Communications, Waltham
King A (1993) The truth about autopoiesis. J Law Soc 20(2):218–236
Koskinen KU (2000) Tacit knowledge as a promoter of project success. Eur J Purch Supply Manage 6:41–47
Koskinen KU (2010a) Autopoietic knowledge systems in project-based companies. Palgrave Macmillan, Houndmills, Basingstoke, Hampshire
Koskinen KU (2010b) Recursive view of the project-based companies' knowledge management. J Knowl Manag 14(2):258–268
Koskinen KU (2013) Knowledge production in organizations: a processual autopoietic view. Springer, Heidelberg
Koskinen KU, Pihlanto P, Vanharanta H (2003) Tacit knowledge acquisition and sharing in a project work context. Int J Proj Manag 21(4):281–290
Lakoff G, Johnson M (1980) Metaphors we live by. University of Chicago Press, Chicago
Lam A (1997) Embedded firms, embedded knowledge: problems of collaboration and knowledge transfer in global cooperative ventures. Organ Stud 18(6):973–996
Landesman C (1997) An introduction to epistemology. Blackwell Publishers, Cambridge
Langley A, Tsoukas H (2010) Introducing 'perspectives on process organization studies'. In: Hernes T, Maitlis S (eds) Process, sensemaking, and organizing. Oxford University Press, Oxford, pp 1–26
Laszlo E (1972) The systems view of the world. George Brazilier, New York
Latour B (1986) Visualization and cognition: thinking with eyes and hands. Knowl Soc: Stud Sociol Cult Past Present 6:1–40
Lehrer K (1990) Theory of knowledge. Routledge, London
Leonard-Barton D (1995) Wellsprings of knowledge. Harvard Business School Press, Boston
Leonard-Barton D, Sensiper S (1998) The role of tacit knowledge in group innovation. Calif Manag Rev 40(3):112–132
Leroy F, Ramanantsoa B (1997) The cognitive and behavioral dimension of organizational learning in a merger: an empirical study. J Manag Stud 34(6):871–894
Levy D (1994) Chaos theory and strategy: theory, application, and managerial implications. Strateg Manag J 15:167–178
Luhmann N (1984) Soziale Systeme. Suhrkamp, Frankfurt am Main
Luhmann N (1986) The autopoiesis of social systems. In: Geyer F, van der Zouwen J (eds) Sociocybernetic paradoxes. Sage, Beverly Hills, pp 172–192
Luhmann N (1988) The unity of the legal system. In: Teubner G (ed) Autopoietic law, a new approach to law and society. Walter de Gruyter, Berlin, pp 12–35
Luhmann N (1990) Soziale Sisteme. Grundriß einer allgemeinen Theorie. Suhrkamp, Frankfurt am Main
Luhmann N (1990a) The autopoiesis of social systems. In: Luhmann N (ed) Essays of self-reference. Columbia University Press, New York, pp 1–20
Luhmann N (1992) Die Wissenschaft der Gesellshaft. Suhrkamp, Frankfurt am Main
Luhmann N (1993) Soziologische Aufklärung 3: Soziales System, Gesellschaht, Organisation. Westdeutscher Verlag, Opladen
Luhmann N (1994) Die Wissenschaft der Gesellschaft. Suhrkamp, Frankfurt am Main
Luhmann N (1995a) Gesellschaftsstruktur und Semantik 4: Studien zur Wissenssoziologie der modernen Gesellschhaht. Suhrkamp, Frankfurt am Main
Luhmann N (1995b) Social systems. Stanford University Press, Stanford

Luhmann N (1997) Die Gesellschaft der Gesellschaft. Suhrkamp, Franfurt am Main
Luhmann N (2000) Organisation und Entscheidung. Westdeutscher Verlag, Opladen
Luhmann N (2013) Introduction to systems theory. Polity Press, Cambridge
Lyles MA, Schwenk CR (1992) Top management, strategy and organizational knowledge structures. J Manag Stud 29:155–174
Maier GW, Prange C, von Rosenstiel L (2001) Psychological perspectives of organizational learning. In: Dierkes M, Berthoin Antal A, Child J, Nonaka I (eds) Handbook of organizational learning and knowledge. Oxford University Press, Oxford, pp 14–34
March JG, Olsen JP (1976) Ambiguity and choice in organizations. Universitetsforlaget, Bergen
March JG, Olsen JP (1989) Rediscovering institutions: the organizational basis of politics. The Free Press, New York
Marquardt MJ (1996) Building the learning organization. McGraw-Hill, New York
Maturana HR (1970) Biology of cognition. In: Maturana HR, Varela FJ (eds) Autopoiesis and cognition: the realization of the living. Reidel, Dordrecht, pp 5–58
Maturana H (1974) Cognitive strategies. In: von Foerster H (ed) Cybernetics of cybernetics. Biological Computer Laboratory, University of Illinois, Illinois
Maturana HR (1975) Communication and representation functions. In: Encyclopedic de la Pleiade, Series Methodique. Gallimard, Paris
Maturana HR (1978) Biology of language: the epistemology of reality. In: Miller GA (ed) Psychology and biology of language and thought: essays in honour of Eric Lenneberg. Academic, New York, pp 27–63
Maturana HR (1980a) Autopoiesis: reproduction, heredity and evolution. In: Zeleny Autopoiesis M (ed) Dissipative structures and spontaneous social orders. Westview Press, Boulder, pp 45–79
Maturana HR (1980b) Introduction. In: Maturana HR, Varela F (eds) Autopoiesis and cognition: the realization in the living. Reidl, Dordrecht, pp xi–xxx
Maturana HR (1988) Reality: the search for objectivity or the quest for a compelling argument. Ir J Psychol 9(1):25–82
Maturana HR (1991) Science and daily life: the ontology of scientific explanation. In: Steier F (ed) Research and reflexivity. Sage, Beverly Hills, pp 30–52
Maturana HR (2002) Autopoiesis, structural coupling and cognition: a history of these and other notions in the biology of cognition. Cybern Human Knowing 9:5–34
Maturana HR, Varela F (1980) Autopoiesis and cognition: the realization of the living. Reidel, Dordrecht
Maturana HR, Varela F (1992) The tree of knowledge. Shambhala, Boston
Maula M (2006) Organizations as learning systems: 'living composition' as an enabling infrastructure (advanced series in management). Elsevier Science, London
Maurer K (2010) Communication and language in Niklas Luhmann's systems-theory. www.fflch.usp.br/dlm/alemao/pandaemoniumgermanicum
Merley-Ponty M (1963) The structure of behaviour. Beacon Press, Boston
Mesle CR (2008) Process-relational philosophy. An introduction to Alfred north Whitehead. Templeton Press, West Conshohocken
Meyer AD (1984) Mingling decision making metaphors. Acad Manag Rev 9:6–17
Mingers J (1995) Self-producing systems: implications and applications of Autopoiesis. Plenum Press, New York
Mingers J (2010) Realising systems thinking: knowledge and action in management science. Springer, New York
Mintzberg H, Waters JA (1990) Studying deciding: an exchange of views between Mintzberg and Waters, Pettirew, and Butler. Organ Stud 11:1–16
Mitleton-Kelly E (2003) Ten principles of complexity and enabling infrastructures. In: Mitleton-Kelly E (ed) Complex systems and evolutionary perspectives on organizations: the application of complexity theory on organizations (advanced series in management). Elsevier Science, London, pp 23–50
Monod J (1974) Change and necessity. Fontana, London

Mooradian N (2005) Tacit knowledge: philosophic roots and role in KM. J Knowl Manag 9 (6):104–113
Moorman C, Miner AS (1997) The impact of organizational memory on new product performance and creativity. J Mark Res 34(1):91–106
Moorman C, Miner AS (1998) Organizational improvisation and organizational memory. Acad Manag Rev 23(4):698–723
Morgan G (1996) Images of organization, 2nd edn. Sage, Thousand Oaks
Morin E (1982) Can we conceive of a science of autonomy? Hum Syst Manag 3:201–306
Nassehi A (2005) Organizations as a decision machines: Niklas Luhmann's theory of organized social systems. In: Jones C, Munro R (eds) Contemporary organization theory. Blackwell Publishing, Oxford, pp 178–191
Nelson RR (1994) Why do firms differ, and how does it matter? In: Rumelt RP, Schendel DE, Teece DJ (eds) Fundamental issues in strategy. Harvard Business School Press, Boston, pp 247–269
Nelson RR (1995) Recent evolutionary theorizing about economic change. J Econ Lit XXXIII:48–90
Nelson RR, Winter SG (1982) An evolutionary theory of economic change. The Belknap Press of Harvard University Press, Cambridge
Newell S, Robertson M, Scarbrough H, Swan J (2002) Managing knowledge work. Palgrave Macmillan, Basingstoke
Nonaka I, Takeuchi H (1995) The knowledge-creating company: how Japanese companies create the dynamics of innovation. Oxford University Press, Oxford
O'Connor J, McDermott I (1997) The art of systems thinking. Essential skills for creativity and problem solving. Thorsons, London
Orr JE (1990) Sharing knowledge, celebrating identity: community memory in a service culture. In: Middleton D, Edwards D (eds) Collective remembering. Sage, London, pp 169–189
Parkin M (2003) Tales for coaching. Kogan Page, London
Pettigrew TF (1997) Generalized intergroup contact effects on prejudice. Personal Soc Psychol Bull 23:173–185
Piaget J (1936) The origins of intelligence in children. W.W. Norton, New York
Pihlanto P (2005) From economic man to the holistic individual. A quest for a realistic notion of the human actor. In: Tengblad S, Solli R, Czarniawska B (eds) The art of science. Liber and Copenhagen Business School Press, Copenhagen, pp 87–110
Pihlanto P (2009) Decision-maker in focus: holistic individual in the theater of consciousness. LAP LAMBERT Academic Publishing AG and Co. KG, Köln
Polanyi M (1958) Personal knowledge. Towards a post-critical philosophy. Routledge, London
Polanyi M (1966) The tacit dimension. Routledge and Kegan Paul, London
Postman L (1976) Methodology of human learning. In: Estes K (ed) Handbook of learning and cognitive processes, Attention and memory, vol 4. Lea Lawrence Erlbaum Associates Publishers, Hillsdale, pp 11–69
Prigogine I (1980) From being to becoming. W.H. Freeman and Company, San Francisco
Raven P, Johnson C (1991) Understanding biology. Mosby-Year Book, St Louis
Reissman CK (1993) Narrative analysis. Sage, Newbury Park
Rescher N (1996) Process metaphysics. An introduction to process philosophy. State University of New York Press, Albany
Rescher N (2000) Process philosophy. A survey of basic issues. University of Pittsburgh Press, Pittsburgh
Rescher N (2004) Process philosophy. In: Zalta EN, Weber M (eds) *After whitehead: rescher on process metaphysics*. Ontos Verlag, Frankfurt am Main
Rorty R (1980) Philosophy and the Mirror of nature. Princeton University Press, Princeton
Rosenberg N (1982) Inside the black box: technology and economics. Cambridge University Press, Cambridge
Russell B (1910–1911) Knowledge by acquaintance and knowledge by description. Proc Aristot Soc 11:108–128

Savage CM (1996) 5th Generation management: co-creating through virtual enterprising. Dynamic teaming, and knowledge networking. Butterworth-Heinemann, Boston
Schein EH (1980) Organizational psychology. Prentice-Hall, New Jersey
Schön DA (1983) The reflective practitioner: how professionals think in action. Basic Books, New York
Schulz M, Hernes T (2013) A temporal perspective on organizational identity. Organ Sci 24 (1):1–21
Schutz A (1967) The phenomenology of the social world. Heineman Educational Books, London
Schutz A (1970) On phenomenology and social relations. University of Chicago Press, Chicago
Seidl D (2005) Organizational identity and self-transformation: an autopoietic perspective. Ashgate Publishing, Hants
Seidl D, Becker KH (2005) Niklas Luhmann and organization studies. Liber and Copenhagen Business School Press, Oslo
Sherburne DW (1966) A key to whitehead's process and reality. The University of Chicago Press, Chicago
Simon HA (1993) Strategy and organizational evolution. Strateg Manag J 14:131–142
Sims HP Jr, Gioia DA (eds) (1986) The thinking organization. Jossey-bass, San Francisco
Singley M, Anderson JR (1989) The transfer of cognitive skill. Harvard University Press, Cambridge
Sole D, Edmondson A (2002) Situated knowledge and learning in disperse teams. Br J Manag 13:17–S34
Spencer-Brown G (1979) The Laws of form. E.P. Dutton, New York
Stacey RD (2001) Complex responsive processes in organizations: learning and knowledge creation. Routledge, New York
Starbuck WH (1992) Learning by knowledge-intensive firms. J Manag Stud 29:713–740
Starbuck WH, Milliken FJ (1988) Challenger: fine tuning the organization until something breaks. J Manag Stud 25(4):319–340
Steup M (2017) Epistemology. In: Zalta EN (ed) The Stanford encyclopedia of philosophy. Metaphysics Research Lab, Stanford University – via Stanford Encyclopedia of Philosophy, Stanford, CA
Taylor C (1985) Human agency and language: philosophical papers 1. Cambridge University Press, New York
Teubner G (1991) Autopoiesis and steering: how politics profit from the normative surplus of capital. In: in't Veld RJ, Schaap L, Terneer CJAM, van Twist MJW (eds) Autopoiesis and configuration theory: new approaches to social steering. Kluwer, Dordrecht, pp 127–134
Thompson E (2004) Life and mind: from autopoiesis to neurophenomenology, a tribute to Francisco Varela. Phenomenol Cogn Sci 3:381–398
Thompson E (2007) Cognitivism. In: Mind in life: biology, phenomenology, and the sciences of mind. Harvard University Press, London
Thurow LC (2003) Fortune Favours the bold: what we must do to build a new and lasting global prosperity. Harper Collins, New York
Tsoukas H (1991) The missing link: a transformational view of metaphors in organizational science. Acad Manag Rev 16(3):566–585
Tsoukas H (1996) The firm as a distributed knowledge system: a constructionist approach. Strateg Manag J 17(Special Issue):11–25
Tsoukas H (2005) Complex knowledge: studies in organizational epistemology. Oxford University Press, New York
Tsoukas H, Chia R (2005) On organizational becoming: rethinking organizational change. In: Tsoukas H (ed) Complex knowledge studies in epistemology. Oxford Press, Melbourne
Tsoukas H, Vladimirou E (2001) What is organizational knowledge? J Manag Stud 38(7):973–993
Tuomi I (1996 December) The firm as a distributed knowledge system: a constructionist approach. Strat Manag J 17:11–26
Ushenko AP (1986) Power and events. University of Chicago Press, Chicago

Usher RS (1989) Locating experience in language: towards a poststructuralist theory of experience. Adult Educ Q 40(1):23–32
Van de Ven AH (1992) Suggestions for studying strategy process: a research note. Strateg Manag J 13(Summer Special Issue):169–188
Vanverstraeten R (2009) The autopoiesis of decisions in school organizations: conditions and consequences. In: Magalhaes R, Sanchez R (eds) Autopoiesis in organization theory and practice. Emerald Group Publishing Limited, Bingley, pp 289–302
Varela FJ (1979) Principles of biological autonomy. North-Holland, Amsterdam
Varela FJ, Maturana HR, Uribe R (1974) Autopoiesis: the organization of living systems, its characterization and a model. Biosystems 5(4):187–196
Varela FJ, Thompson E, Rosch E (1991) Embodied mind: cognitive science and human experience. MIT Press, Cambridge
Venzin M, von Krogh G, Roos J (1998) Future research into knowledge management. In: von Krogh G, Roos J, Kleine D (eds) Knowing in firm: understanding, managing, and measuring knowledge. Sage, London, pp 26–66
Vicari S, Troilo G (1999) Organizational creativity: a new perspective from cognitive systems theory. In: von Krogh G, Nonaka I, Nishiguchi T (eds) Knowledge creation: a source of value. Macmillan, London, pp 63–88
Von Bertalanffy L (1969) General system theory – foundations, development, applications. George Braziller, New York
Von Foerster H (1972) Responsibilities of competence. J Cybern 282:1–6
Von Foerster H (1984) On constructing a reality. In: Watzlawick P (ed) The invented reality. Norton, New York, pp 41–61
Von Foerster H (2003) On constructing a reality. In: Understanding: essays on cybernetics and cognition. Springer, New York, pp 211–227
Von Krogh G, Roos J (1995a) Organizational epistemology. St Martin's Press, New York
Von Krogh G, Roos J (1995b) Conversation management. Eur Manag J 13(4):390–394
Von Krogh G, Roos J, Slocum K (1996) An essay on corporate epistemology. In: von Krogh G, Roos J (eds) Managing knowledge: perspectives on cooperation and competition. Sage, London, pp 157–183
Weber CFM (2004) After Whitehead: Rescher on process metaphysics. Frankfurt am Main, Ontos Verlag
Weick KE (1969) The social psychology of organizing. Addison-Wesley, Reading
Weick KE (1979) The social psychology of organizing. Random House, New York
Weick KE (1995) Sensemaking in organizations. Sage, Thousand Oaks
Weick K, Sutcliffe KM, Obstfeld D (2005) Organizing and the process of sensemaking. Organ Sci 16(4):409–421
Weinberg GM (2001) An introduction to general systems thinking: silver, Anniversary edn. Dorset House Publishing, New York
Wenger E (1991) Communities of practice: where learning happens. In: Benchmark. Dorset House Publishing, New York, pp 82–84
Wenger E (1998) Communities of practice. Cambridge University Press, Cambridge
Whitehead AN (1938) Models of thought. Free Press, New York
Whitehead AN (1978) Process and reality. Free Press, New York
Wiener N (1948) Cybernetics: or communication and control in the animal and the machine. MIT Press, Cambridge
Wikström S, Normann R et al (1994) Knowledge and value. Routledge, London
Winograd T, Flores F (1986) Understanding computers and cognition. Addison-Wesley, Reading
Wittgenstein L (2009) Philosophical investigations. Wiley-Blackwell, New York
Ylijoki OH (2005) Academic nostalgia: a narrative approach to academic work. Hum Relat 58 (5):555–575
Zalta EN (2015) The Stanford encyclopedia of philosophy. The Metaphysics Research Lab, Stanford, CA
Zimmerman B, Hurst DK (1993) Breaking the boundaries: the fractal organization. J Manag Inq 2 (4):334–355

The manufacturer's authorised representative in the EU is Springer Nature Customer Service Centre GmbH, Europaplatz 3, 69115 Heidelberg, Germany. If you have any concerns regarding our products, please contact ProductSafety@springernature.com

Printed and bound by CPI Group (UK) Ltd, Croydon, CR0 4YY
30/09/2026
02225468-0012